Oscilloscopes: A Manual for Students, Engineers, and Scientists

David Herres

Oscilloscopes: A Manual for Students, Engineers, and Scientists

 Springer

David Herres
Private Consultant
Clarksville, NH, USA

ISBN 978-3-030-53884-2   ISBN 978-3-030-53885-9   (eBook)
https://doi.org/10.1007/978-3-030-53885-9

This Springer imprint is published by the registered company Springer Nature Switzerland AG
The registered company address is: Gewerbestrasse 11, 6330 Cham, Switzerland

Contents

Introduction

Introductions serve a variety of purposes. First, the reader will want to know what background is presupposed. This book, a treatise on oscilloscope operation including the display, measurement, and analysis of waveforms, will be most useful for readers with some theoretical and practical knowledge of electronics. A complete mastery is not required or expected and, in fact, is not possible given the vastness of the field.

The same comment applies in another very large area—mathematics. You should come to this book with the ability to work with algebraic equations as they relate to the ordinary (non-quantum, non-relativistic) world. You need to know about the fundamental trigonometric functions pertaining to a right triangle inscribed within a circle, and some knowledge of calculus is good, but not essential. The aspiration in writing this book is to provide tools and insights for students, technicians, and experienced engineers who want to look a bit deeper into the oscilloscope's properties, functionality, and practical applications.

There are a number of manufacturers who offer high-quality, well-designed oscilloscopes that are a joy to work with. Of necessity, I have used a single model, the Tektronix MDO3104, in many demonstrations herein. Many readers will own or have access to one or more different oscilloscopes. For the most part, you will not have a problem applying procedures laid out here to these competing makes. It is like going from a Windows to a Mac computer. There is some different terminology and layering of menus in competing oscilloscopes, but the waveforms are the same as are our techniques for analyzing them.

Most manufacturers' websites offer excellent documentation. Invariably, you can download free of charge user and programming manuals. This is a good approach when you need to translate knowledge applicable to competing instruments.

In addition to the foregoing, the Introduction is a good place to offer a few practical suggestions with regard to safety. Because the oscilloscope is an electrical instrument, there is a potential to confront the twin demons of electrical shock and fire. These dangers can be catastrophic, but with reasonable precautions and the basic training that should precede work in the lab, shop, or on the factory floor, you should be able to protect yourself, coworkers, and passersby.

The user manual that accompanies even the simplest electronic equipment contains in the first few pages abundant cautions and warnings. Of course an ordinary bench-type oscilloscope should not be operated or stored in a damp location. The power cord should be plugged into a receptacle that supplies the correct voltage and has over-current protection. For this to be functional, it is essential to resist any impulse to saw off the ground plug since it provides equipment-grounding continuity back to the premises electrical service and ground electrode.

At the analog channel inputs voltage limits should be observed, typically 300 volts at 60 Hz and less at higher frequencies. The amount is stamped on the oscilloscope front panel and the user will find details in the manual.

The risk in belaboring these and other cautionary items is that some readers may be put off, particularly those who would benefit most. There is one aspect of oscilloscope operation that bears repeating because if it is neglected there is the potential for smoke, sparks, and far worse. If a certain improper connection is made, intense fault current will flow through a part of the circuit under test, the probe, and the oscilloscope. To be succinct, the probe ground return lead must never be connected to a terminal or conductor that floats above but is referenced to ground potential. Within the oscilloscope, there is a low-impedance current path between the probe ground return lead and, through the power cord, the grounding conductor in the premises wiring, which ultimately leads back to the grounding system at the electrical service. A high-level fault current will result even if the probe tip is not contacting a part of the circuit and indeed if the oscilloscope is not even powered up.

A way to deal with this problem is to saw off the grounding conductor at the power plug. The problem with that approach is that the user is disabling a very important safety feature common to all tools that are not double-insulated. If a ground fault occurs within the instrument or the circuit under investigation, the grounding conductor ensures that the circuit breaker or fuse will open the circuit, preventing possibly lethal shock to the user.

The floating ground hazard can be eliminated by using a differential probe, but the problem here is the very high expense of that item and also that the bandwidth of the oscilloscope is severely compromised.

Many technicians and engineers who work on equipment that contains floating voltages, notably three-phase variable-frequency motor drives and switching power supplies, use hand-held, battery-powered oscilloscopes that have channel inputs isolated from ground and insulated from one another. But don't assume anything. Some of these instruments do not have isolated channel inputs. Consult the manufacturer if the manual is not explicit in this regard.

A final thought: Quantum mechanics asserts that the smallest particles exhibit wave properties. These oscillations are additive, so that the entire universe is a single, vast, unbelievably complex waveform. Besides being a measuring instrument and analytic tool for astrophysicists, automotive engineers, and theoreticians, the oscilloscope is a window to the universe. In this book we'll work on some practical applications and along the way endeavor to put it all in a larger perspective.

Chapter 1
Waveforms and Instrumentation Overview

Abstract The oscilloscope emerged from a relatively primitive beginning when electrical experimenters built oscillographs that would extract waveforms from acoustic and electrical phenomena and record them on rotating paper cylinders. Frequency response and bandwidth were severely limited by inertia of the pen and ink recording equipment. Various alternative strategies involved vibrating mirrors that would project light images and later the cathode-ray tube.

In the technological boom that followed World War II, oscilloscopes were manufactured in great numbers and became available not only in the laboratory but in technician's shops and for advanced home experimenters. Digital and flat-screen versions flourished and have endured.

Oscilloscopes and other equipment included spectrum analyzers in the laboratory and for advanced fieldwork. Techniques employed by the first generation of TV technicians have evolved into far more useful methods for analyzing waveforms drawing conclusion that pertain to the specific project.

Charles Fourier's mathematical transformation became applicable to electrical phenomena. With the introduction of the fast Fourier transform (FFT), it became an integral part of the contemporary oscilloscope. FFT functionality clarifies circuit behavior and facilitates fault diagnosis when existing equipment or new designs fail to perform as expected.

In the world of electronic test equipment, the oscilloscope and its coworker, the spectrum analyzer, occupy the high ground. In contrast to the digital multimeter, itself a highly versatile and competent instrument, these machines permit the user to go farther, visualizing AC waveforms in exquisite detail rather than merely quantifying them in terms of amplitude and frequency.

In this book we'll examine the many ways in which engineers, technicians, and students use these wonderful instruments to see electrical and electromagnetic energy in real time, as a static image when the information flow is paused or in retrospect as is possible in a digital storage oscilloscope.

We'll also look under the hood to see in detail how the oscilloscope captures, processes, and displays signals, and we'll also have a lot to say about techniques and procedures for signal tracing within electronic equipment and about analyzing

© Springer Nature Switzerland AG 2020
D. Herres, *Oscilloscopes: A Manual for Students, Engineers, and Scientists*,
https://doi.org/10.1007/978-3-030-53885-9_1

analog and digital signals when the equipment under examination is in failure mode in the repair shop or it is in some stage of development as a provisional prototype.

To begin, we'll discuss the nature of wave motion, both human-made and as it existed in the universe long before humans emerged and acquired the theoretical background and experimental expertise to indulge in such inquiries.

Wave vs. Waveform

A waveform corresponds to a changing or as a limiting case static state of affairs as in the horizontal line representing an unchanging DC voltage.

It can represent a varying value in the real world or a mathematical abstraction such as a trigonometric function that assumes different values, corresponding to an angle denoted by Greek theta (θ) varying in time. In all cases, we must be aware that the waveform in an oscilloscope display is not the actual wave but rather a graph of that wave. (It would be improper to speak of a waveform as a physical entity. It is the wave that moves through space, and the waveform as displayed in an oscilloscope represents it.) In its most fundamental form, in the time domain, the waveform is an image of a moving point whose position in two-dimensional space relates to its value on a Y-axis denoting amplitude, typically in volts, and to its value on an X-axis denoting time, typically in seconds. These two axes are by convention arranged to intersect at right angles at the center of an oscilloscope display or of a printed page in a textbook discussion (Gibilisco 2012). Figures 1.1, 1.2, and 1.3 show how this same wave, in this case a 60 Hz sine wave in corresponding to an

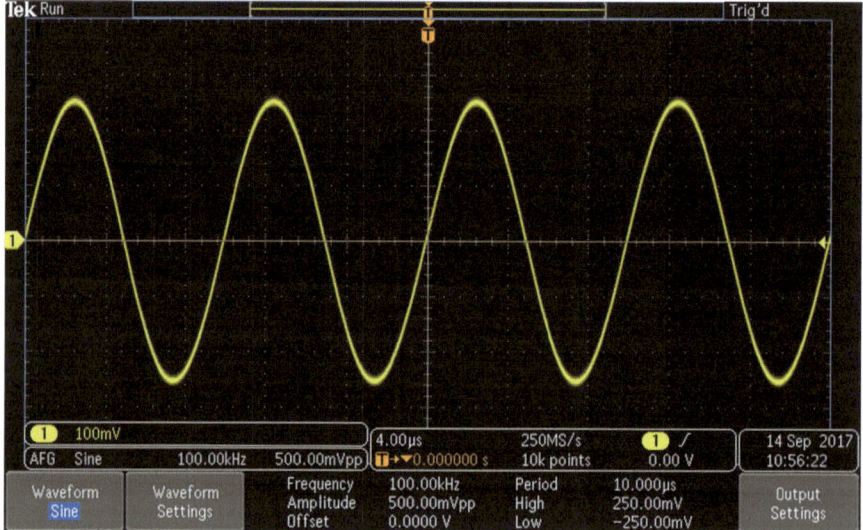

Fig. 1.1 100.00 kHz sine wave in time domain, default scale. (Author's screenshot)

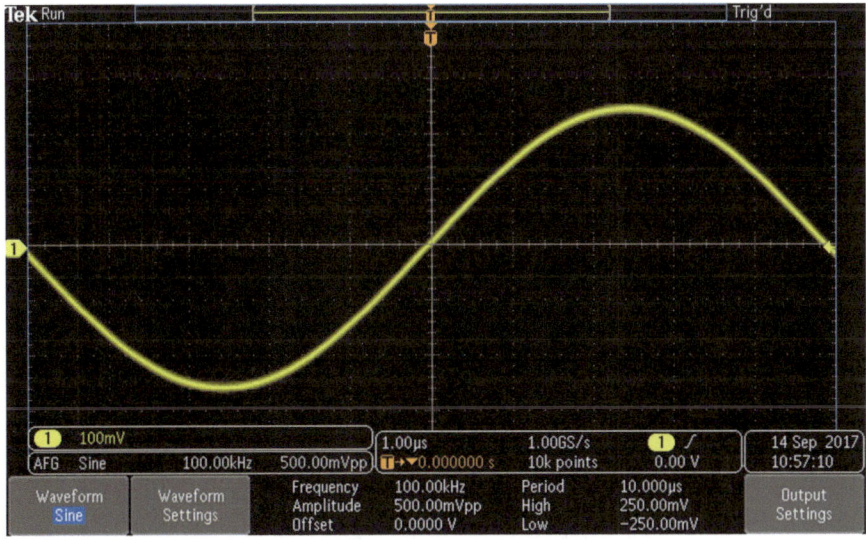

Fig. 1.2 100.00 kHz sine wave in time domain, coarse scale. (Author's screenshot)

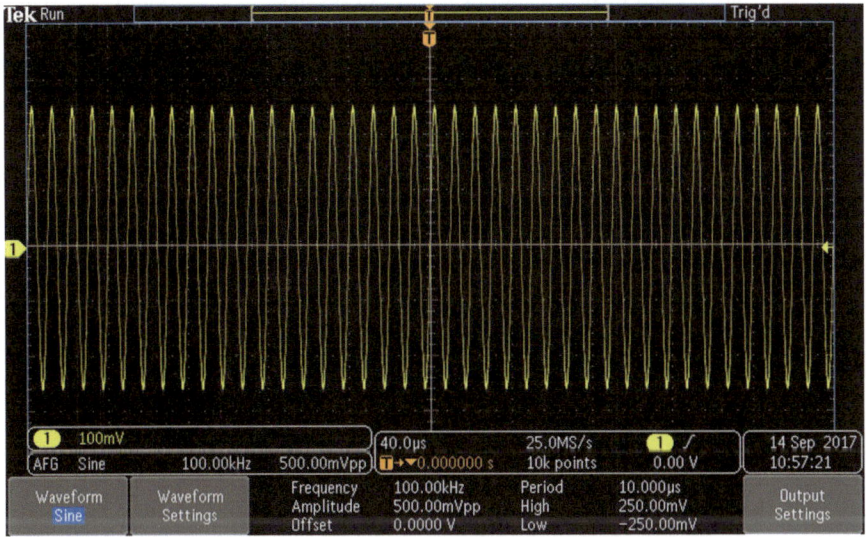

Fig. 1.3 100.00 kHz in time domain, fine scale. (Author's screenshot)

oscillating utility secondary voltage, shown in the time domain, can have radically different appearances depending on the scaling of the two axes, which can be varied simply by turning the horizontal scale control on the front panel of the oscilloscope.

This same signal with a totally different appearance can be displayed in the frequency domain where the X- and Y-axes are defined differently. Here the X-axis is made to represent a range of frequencies, usually with the fundamental situated at the left side or the center of the display. The Y-axis, as in the time domain, represents various levels of amplitude, but now it is power in decibels on a logarithmic scale (for readability) rather than voltage that is shown (Hickman 2001).

Here again, the display can assume radically different appearances, depending on the scaling, as shown in Figs. 1.4, 1.5, and 1.6.

A third view of the same signal with yet another much different appearance is the spectrogram, as shown in Fig. 1.7, where frequency is laid out along the X-axis as in the frequency domain, but time is represented on the Y-axis so that the function along with any changes appears to be moving upward continuously, in a very stately fashion. Since it would be unwieldy and counterproductive to build an oscilloscope with three axes intersecting at right angles, amplitude is shown by varying colors in the display. The color scheme can be altered or reversed in the instrument preferences, to suit the user.

I'll have much more to say about various types of waveforms in the chapters that follow. Also we'll see how they can be viewed using the many tools and operating modes built into today's amazing oscilloscopes. For now, the point is that the actual electrical or electromagnetic wave is not the same as the graph of it that we see in the oscilloscope display. A real-world, real-time wave is a pulsating entity that is moving (or stationary, as in a standing wave, but located) in space and time. What we see in the oscilloscope display is a graphical representation of the waveform, derived for our benefit from the wave but in its own way equally miraculous, a testament to our present-day technological prowess. The oscilloscope in a real sense is a window on the universe.

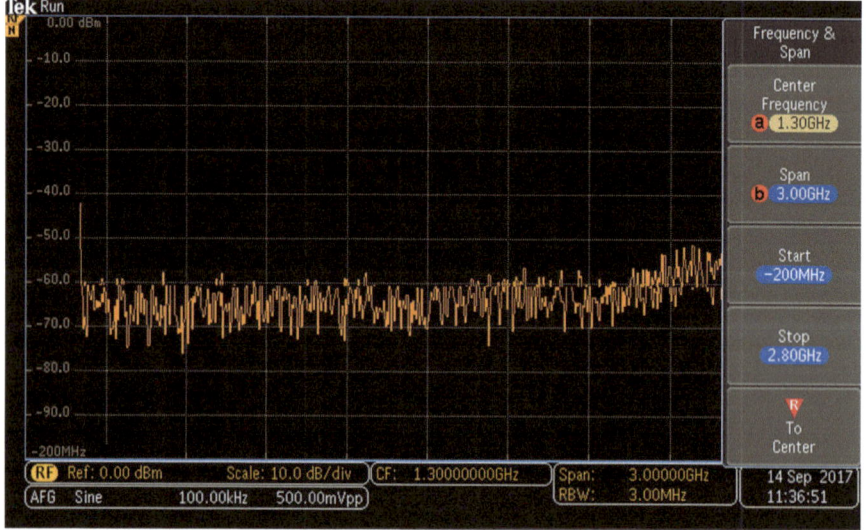

Fig. 1.4 Sine wave in frequency domain, center frequency 1.3 GHz. (Author's screenshot)

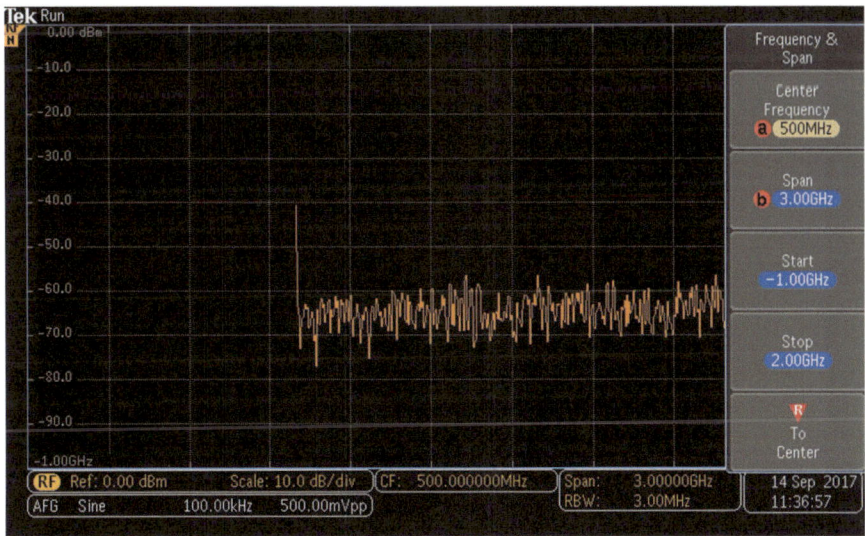

Fig. 1.5 Sine wave in frequency domain, center frequency 500 MHz. (Author's screenshot)

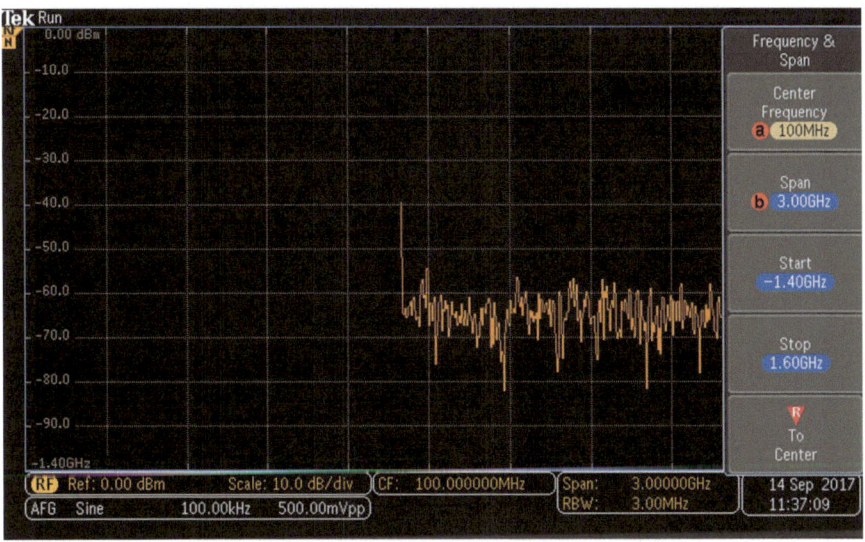

Fig. 1.6 Sine wave in frequency domain, center frequency 100 MH. (Author's screenshot)

Before examining ways in which the oscilloscope and its highly developed relative, the spectrum analyzer, are used in a contemporary lab or repair shop, we'll look at some waveform basics, and also we'll trace the interesting history of this profound instrument.

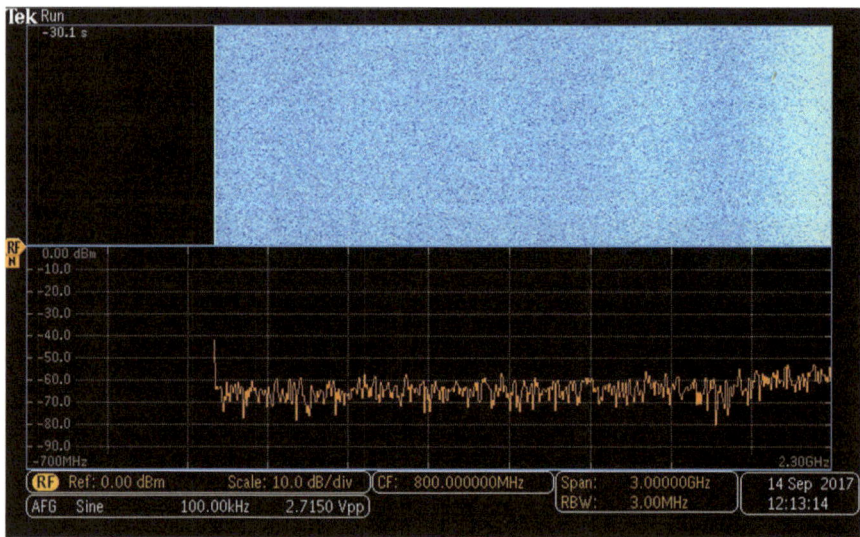

Fig. 1.7 Spectrogram of sine wave. The bright vertical line at the left corresponds to the fundamental. The speckled appearance of the background corresponds to the noise floor in the instrument, which can also be seen as irregularities in the trace, outside of the fundamental. (Author's screenshot)

Waveform Fundamentals

Wave action takes many forms, any of which can be imaged in an oscilloscope. The instrument is designed to receive at its inputs electrical energy and to display it, nowadays on a flat screen rather than cathode-ray tube.

Nonelectrical waveforms can be accessed – it's just a matter of deploying appropriate sensors and transducers to translate temperature, water pressure, wind speed, or other parameters into volts.

There are three fundamental types of waves – mechanical, electrical, and electromagnetic. Mechanical waves consist of an oscillation in some local medium. A familiar example is the acoustic wave, an oscillation that we perceive as sound. Ancient humans were well aware of this phenomenon and had a good understanding of its nature, although there are some subtleties that have only recently surfaced.

Acoustic waves as we experience them propagate primarily through air, although any material can function as the medium provided it is sufficiently but not excessively elastic.

Electrical waves consist of oscillations in the flow of particles, generally electrons, through an electrically conductive material, most typically metal. The medium can be drawn to form a long wire, making it useful in power transmission and communication. In the popular imagination, electrons enter a conductor at the input and travel at the speed of light to the output and then traverse some connected load whereupon they travel back to the power source, completing the circuit via a return conductor.

This picture is at best partially accurate. The reality is that the electrons that exit the conductor in a given time segment are not necessarily the same particles that enter. Typically, electrons travel a short distance through a conductor, and some fraction of them encounter an atom. The collision may result in capture of the electron and emission of a different electron. It is like a passenger train, where some individuals disembark at various stops, replaced by others, and some passengers travel straight through to the destination.

Notwithstanding, this electrical current, if AC, as well as the voltage that is applied to the conductors at the input, exhibits wave oscillations, simple or complex, that can be measured, recorded, or shown in an oscilloscope display.

Oscillating electrical current as well as static DC can flow also through electrical devices such as active semiconductors, which require an external bias, and also through passive components such as resistors, capacitors, and inductors. Some of these, essentially resistors, consume power, dissipating it in the form of heat. Others may seem to consume power, but actually they store it – capacitors in an electrostatic field between the plates, and inductors in a magnetic field within and outside the conductor. In all cases, electrical current in amperes or electromagnetic force in volts can be displayed in an oscilloscope, which exhibits a detailed image of the waveform, if it is AC or a complex oscillating signal, along with relevant numerical metrics. Engineers, technicians, and students make abundant use of this information, which paints a picture of what is really going on in an electrical circuit.

Energy can also propagate in the form of electromagnetic waves traveling through free space, apparently with no material medium, or through matter that is transparent to it, based on the frequency of the wave, and the spacing of the particles that comprise the medium.

Wave propagation through a vacuum with no detectable material medium has been highly problematic in the history of science as well as for most of us in our evolving understanding of this complex issue. The problem became acute in nineteenth-century physics, becoming only partially resolved first in Einstein's two relativity theories and then in twentieth-century quantum mechanics.

Before James Clerk Maxwell's time, the speed of light had been measured to a fair degree of accuracy, and since 1831 it was known that magnetism could produce electricity. Maxwell now saw that his equations for electromagnetic propagation had wave-like solutions. He succeeded in calculating the speed of those waves, which depended upon lab experiments. The speed was 310,740,000 meters per second, which happened to be the speed of light! Maxwell concluded that light was indeed a form of electromagnetic radiation.

Maxwell and his colleagues assumed that light and other electromagnetic radiation would consist of vibrations in some luminiferous ("light-bearing") aether. They knew that light resembles sound, which since ancient times had been known to consist of vibrations in air and other media. This aether would be either stationary or it would move through space at a constant speed and always in the same direction within its spatial context.

They also knew that the earth traveled rapidly through space. Its motion was comprised of its revolution about the sun and the sun's absolute motion. Because the earth rotates about an axis, that motion also played a role. The idea at the time was that the uniform speed of light moving between two points located along the direction of motion was constant, so the time required for the light to travel in the two opposite directions should differ. This would be measureable and analogous to a perceived wind speed caused by the observer's motion with respect to the luminiferous aether.

Albert Michelson (theoretician, planner, conceptualizer) and Edward Morley (technician, mechanic, skilled builder of instruments), both quintessential nineteenth-century American researchers, designed and carried out experiments to precisely measure the aether wind and thus verify its existence.

At the time it was widely held that the Michelson-Morley experiments consisted of these experimenters on separate mountain tops, perhaps 20 miles apart, measuring the difference in transit times between simultaneous light beams going both ways. But actually, measurement of the speed of light at this time was not sufficiently accurate to perform this sort of experiment. As a viable alternative, Michelson designed and Morley built an interferometer in 1887. It was located within the confines of a single room. This ingenious instrument consisted of a light source, half-silvered glass plate, mirrors, and optics. It was mounted on a heavy stone slab that floated in a pool of mercury in order to isolate it from any random motion or temperature changes. It could be rotated to alter the orientation of the instrument with respect to the earth's linear motion.

A beam of light was split into two parts. One part was reflected by the half-silvered mirror and the other transmitted through it. These separate beams traveled to mirrors mounted at the ends of two perpendicular arms. Then they were reflected back, recombining in an eyepiece. If there were the anticipated difference in transit times, constructive and destructive interference between the out-of-phase light waves would cause visible fringing. The amount of fringing would vary as the apparatus was rotated in the pool of mercury and its orientation with respect to earth's presumed absolute motion was altered.

Michelson and Morley were astonished to observe that the change in the amount of interference between the two beams of light was far less than anticipated. They repeated the procedure with increasingly accurate and vibration-proof instrumentation, and other experimenters made similar attempts. In 1887, Michelson wrote:

"The experiments on the relative motion of the earth and aether have been completed and the result decidedly negative. The expected deviation of the interference fringes from the zero should have been 0.40 of a fringe – the maximum displacement was 0.02 and the average much less than 0.01 – and then not in the right place. As displacement is proportional to squares of the relative velocities it follows that if the aether does slip past the relative velocity, it is less than one sixth of the earth's velocity."

Many explanations were proposed, such as the idea that the earth was somehow dragging the luminiferous aether along with it, though this would appear impossible since the same phenomena should be observable at other locations. An alternate explanation was the proposed Lorentz-Fitzgerald contraction (1889). It asserted that the Michelson-Morley anomaly was in reality due to the shortening of any moving object along its direction of motion.

Eventually Albert Einstein resolved the Michelson-Morley contradiction by asserting:

- The laws of physics are the same for all observers in uniform motion relative to one another.
- The speed of light in a vacuum is the same for all observers, regardless of their relative motion. This postulate definitively accounts for the Michelson-Morley null result.

The bottom line is that while mechanical waves require a medium to move from one location to another, electromagnetic waves propagate through a vacuum with no known material medium. When the initial disturbance occurs, whether a small LED or star, many orders of magnitude larger, forming in space, a field is established, and this is necessary for the electromagnetic waves to propagate. But what precisely is a field? Perhaps it is just a conceptual thing, conceived by us to explain electromagnetic propagation through empty space. On the other hand, if there are vibrations, surely there must be something, immaterial but nonetheless real, that is vibrating.

A field has been variously defined as follows:

- An electromagnetic field is a special kind of substance by which charged particles or physical bodies with a magnetic moment interact.
- A field is a function that returns a value for a point in space.

The first definition uses the word substance but appears to hedge in using the phrase "special kind of..." as if a field is a substance, which by all accounts is material, and yet it is probably not material and therefore not a substance in the usual sense. We are no closer to understanding the real nature of a field.

The second definition is a little more succinct, and yet it too provides a minimal amount of information regarding the actual nature of a field. Perhaps we can gain some understanding of a field by reviewing what is known:

1. Material is a substance that has a nonzero value of mass, so in this sense, a field appears to be immaterial in contrast to the physical media (metal wire, violin string, air) through which mechanical waves travel or along which standing waves are situated.
2. Electrical energy travels through conductive media, and when it is AC, the waves oscillate at a uniform or varying frequency. In this sense, electrical waves are mechanical.
3. Electromagnetic waves travel through space, either having material content or being devoid of matter, i.e., a vacuum, also at a uniform or varying frequency.
4. Electrical wave energy traveling through a conductor creates a magnetic field around the conductor. Depending upon the amplitude in volts and the frequency in hertz, the conductor can function as a transmitting antenna, emitting electromagnetic energy of the same frequency and a dependent amplitude. Rather than a material medium, the electromagnetic energy traverses an immaterial field.

When and if it impinges upon another conductive body (a receiving antenna), electrical energy will emerge, its frequency or frequencies the same as those of the electromagnetic energy that traversed space and the electrical energy in the original conductor some distance away, with an amplitude that is related although not the same. This, of course, is the basis for radio transmission.

5. The field that conveys the electromagnetic waves is functionally similar to a gravitational field, and also it is functionally similar to electrostatic and magnetic fields, although there are profound differences in how they behave. It should be noted that a body can be subject simultaneously to more than one field, as in the case of a permanent magnet that weighs two pounds.

6. It seems self-evident although difficult to prove that these fields arise when the energy is released at the origin. The fields probably do not exist prior to that event.

7. Electrical waves can be displayed as waveforms in an oscilloscope, merely by applying the associated voltage at an analog input. Electromagnetic waves, by means of a receiving antenna that converts them to oscillating electrical energy, can be similarly displayed. Mechanical waves, by means of appropriate sensors and transducers, can also be converted to electrical waves and be displayed in an oscilloscope, an example being sound waves in air applied to a microphone.

(If you want to see your voice in an oscilloscope display, use either an electret microphone, which generates electrical power in response to oscillations in air pressure, or use a carbon microphone in series with a DC electrical source.)

Today's digital storage oscilloscopes have evolved to become intricate, highly functional instruments with extensive measuring and analytic capabilities that go way beyond merely displaying waveforms of electrical signals connected at the inputs. To fully appreciate contemporary instrumentation in electronic laboratories and advanced repair shops, it is useful to review the early history and subsequent development of oscilloscope technology. Its roots lie in the emergence in the nineteenth century of the oscillograph, which began as a tool to facilitate hand-drawn charts depicting electrical fluctuations.

In its first implementation, graphic artists spent days plotting galvanometer readings taken at close intervals about the circumference of a generator's rotor that was slowly advanced by hand a degree or so at a time, as shown in Fig. 1.8. On graph paper, a curve was drawn to simulate the waveform of the generator's anticipated output with the rotor spinning at rated speed. The waveform was only a rough approximation because it was synthesized in the course of thousands of wave cycles, but this was the beginning of waveform imaging.

The development of automated oscillographs resulted in improved waveform imaging, but still the slow reaction time of mechanical components due to inertia of rest meant that the waveform image had to be compiled over many cycles, combining small segments from successive waveforms as opposed to drawing individual waveform images. These composites were averaged over hundreds, rather than thousands of still images as in the hand-drawn oscillograms, so far greater accuracy became feasible.

Fig. 1.8 This early automated waveform imager incorporated a pen and paper drum to record a waveform built up over many cycles, using a synchronous motor and permanent magnet galvanometer. (Wikipedia)

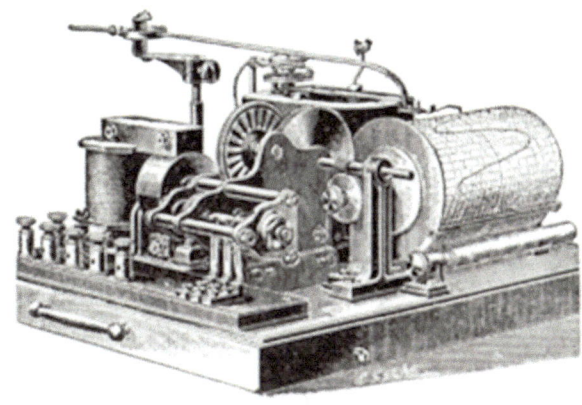

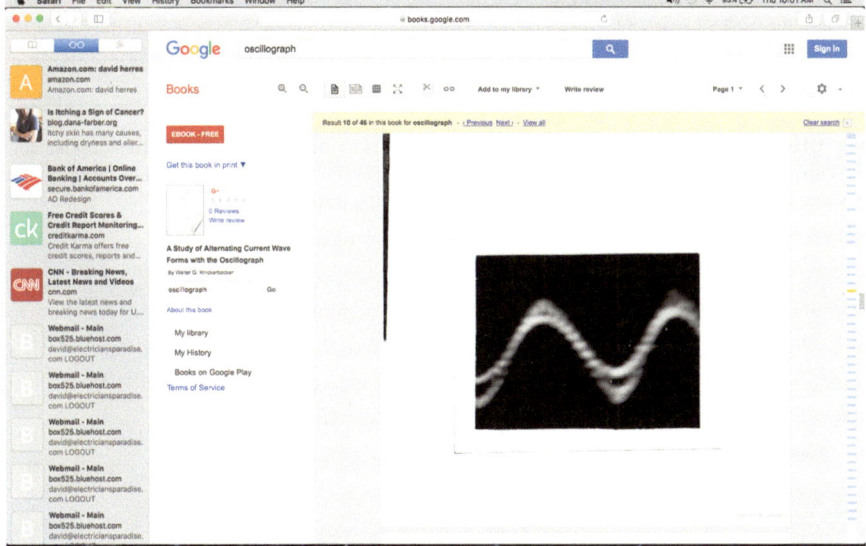

Fig. 1.9 Current and voltage relations in a circuit containing resistance only. It shows that the current and voltage are in phase, and the machine gives a sine curve

A decisive advance was the moving-coil oscillograph, also known as the mirror galvanometer. The pen, paper, and moving drum powered by a synchronous motor were now replaced by a small mirror that vibrated rapidly to draw a line composed of light on a continuous roll of motion picture film. This lightweight assembly could image moderate frequency waveforms in real time.

Experimenters built various increasingly lightweight assemblies, boosting the frequency response (as we now say "bandwidth") of these still primitive instruments. They were capable of displaying more intricate waveforms, as Figs. 1.9, 1.10, 1.11, 1.12, and 1.13, photos taken from an early instrument in 1917, demonstrate.

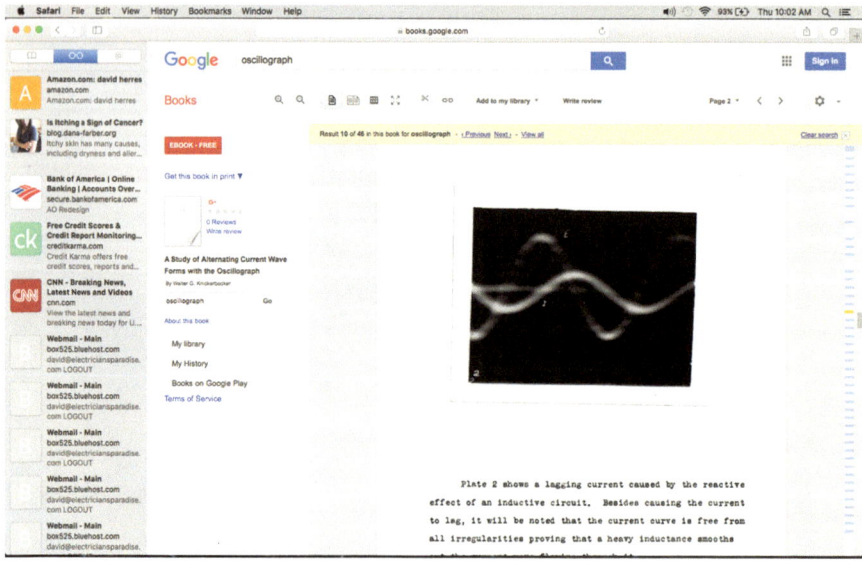

Fig. 1.10 A lagging current caused by the reactive effect of the inductive circuit. Besides causing the current to lag, it will be noted that the current curve is free from all irregularities proving that a heavy inductance smoothes out the current wave flowing through it

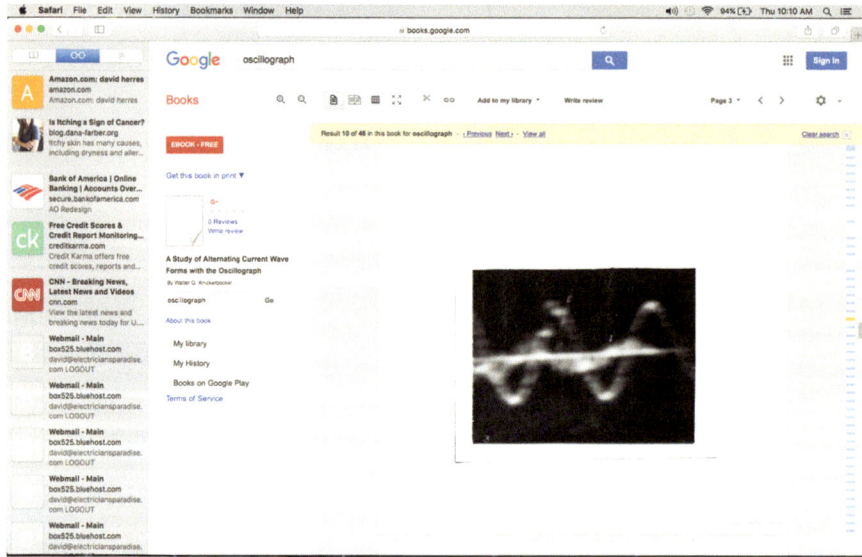

Fig. 1.11 The effect of applying the same voltage as was used in the first two cases to a circuit containing capacitors. It shows a current wave leading the voltage wave about 90 degrees. It also demonstrates the fact that a capacitor tends to emphasize all irregularities in the current wave in a circuit containing resistance, inductance, and capacity in parallel

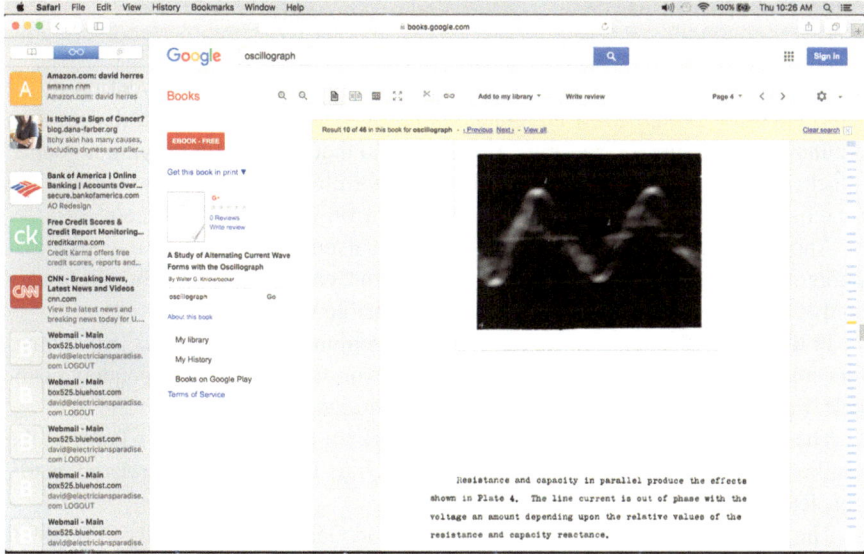

Fig. 1.12 Resistance and capacity in parallel. The line current is out of phase with the voltage an amount depending upon the relative values of the resistance and capacitive reactance

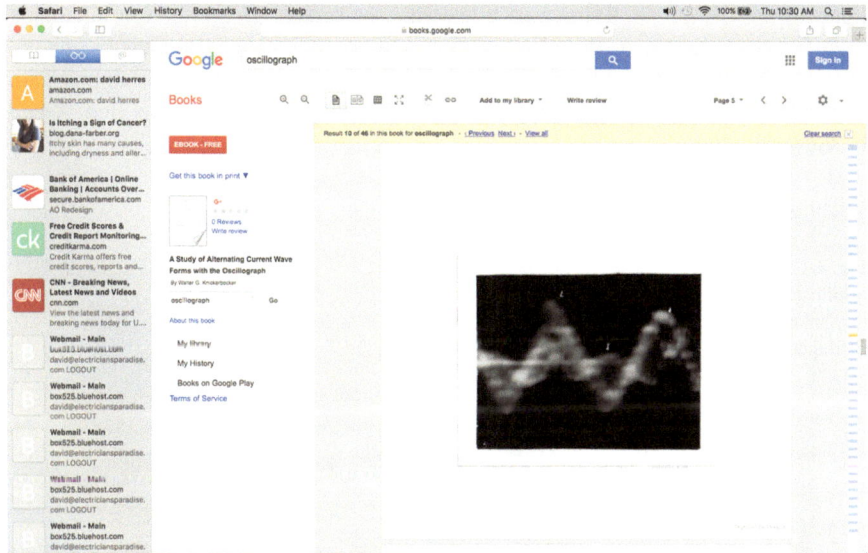

Fig. 1.13 Currents in a circuit containing resistance, inductance, and capacitance in parallel are added algebraically; the resultant is as shown. The indistinct outline of the curve is caused by the rapid shifting of the waveform in circuits of this nature

New Developments

The oscilloscope as we know it began to take form following the invention of the CRT, but prior to that, a few incremental improvements appeared. A little less than a hundred years ago, a small mirror, mounted so that it could tilt varying amounts, attached to a diaphragm on a horn, projected waveforms onto a white screen. The time base, consisting of a rotating polygonal mirror, was unsynchronized.

Another roughly contemporaneous effort involved sound directed toward a diaphragm attached to the gas feed supplying a flame, causing the flame height to vary so that a rotating mirrored polygon produced visible waveforms.

In the mid-1900s, photo-sensitive paper in conjunction with mirror galvanometers created print records of multiple waveforms on separate and independent channels. Frequency response extended into the audio range.

The invention of the CRT laid the groundwork for the instrument we know as the oscilloscope, although in the oscilloscope as well as TV it has been rendered nearly obsolete by our now familiar flat screen.

The CRT, developed from earlier Crookes and Geissler tubes, emerged in 1897. This innovation predated semiconductors, appearing shortly after ac motors and distribution systems (Fig. 1.14).

As opposed to earlier cold-cathode tubes, in the CRT low-voltage electrical energy is applied to a filament, which glows emitting radiation across a wide spectrum. A metallic cathode, situated nearby and heated by the filament, is biased so that it emits a continuous stream of electrons that travel at a fraction of the speed of light toward the anode, flying past it to strike the screen. This radiation inhabits a high-frequency band, in the ultraviolet neighborhood, so it is not visible to the human eye. It is only when the electrons strike the glass screen or its phosphor-coated inner surface that visible light passes through the glass so that it can be viewed.

Fig. 1.14 Early oscilloscopes used CRT's to display waveforms. (Wikipedia)

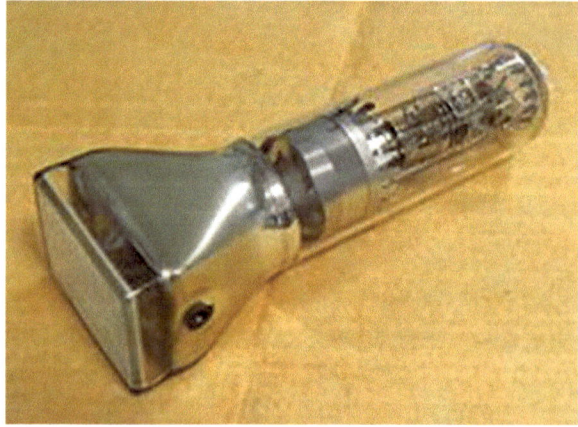

After it is accelerated and focused to make a narrow beam, the stream of electrons is vertically and horizontally deflected in order to create an image or trace on the screen. These two simultaneous deflections are accomplished by deflection coils in a TV or computer monitor and by deflection plates in an oscilloscope. (A viable if very limited oscilloscope can be made from an old black and white TV, feeding the signal into the vertical deflection coil to show amplitude and into the horizontal deflection coil to create a time base).

The incandescent light bulb, as developed by Thomas Edison and his army of technicians in the mid-nineteenth century, consisted of a glass envelope with air evacuated so that the filament would not burn up. Glass blowing was a highly developed art at the time, and it was not difficult to produce bulbs containing a vacuum, with metal conductors entering the envelope, the molten glass making airtight seals around them.

Other experimenters contemplated variations. Electrodes could be situated within a glass bulb containing fractional atmospheres of various gasses. Unconnected by a filament, these electrodes could be energized and the results recorded.

One researcher, Heinrich Geissler, built the first gas discharge bulb in 1857.

In addition to his background in physics, Geissler was an experienced mechanic and skilled glassblower. After working at various institutions, he established a laboratory at the University of Bonn, where he built a hand crank mercury vacuum pump. His glass tubes containing intense vacuum could withstand atmospheric pressure. The Geissler tube, equipped with anode and cathode for positive and negative, when energized emitted a gentle glow. The color depended upon the type of gas that had been introduced into the tube, in all cases at low pressure so as not to interfere with what would later be found to be electrons traveling from cathode to anode. The Geissler tube eventually evolved into our neon bulb. Other gasses that were used included argon and mercury vapor. Experimenters also investigated ionizable substances and metals, especially sodium, and each of these produced a distinctive color. Novelty gadgets and educational devices proliferated. Lecturers demonstrated gas ionization in the Geissler tube and test equipment such as a noncontact high-voltage indicator became available.

The Crookes tube, a cold-cathode device invented around 1875, was a variation that eventually eclipsed in importance the Geissler tube. It revealed to researchers the existence of cathode rays, which were subsequently identified as subatomic particles and called electrons.

In the Crookes tube, high voltage was applied to the cathode, located at the narrow end, and to the anode, connected to a terminal that penetrated the tube at the side. Electromagnetic energy, later characterized as simultaneously particle and wave, traveled from cathode to anode, and a portion of this energy went flying past the anode to strike the large end of the tube, passing through the glass in the form of light. Experimenters mounted a metal shield in the path of the electron flow, and it cast a shadow that appeared at the wide end of the tube. The shield, in the shape of a Maltese cross or other symbol, could be mounted on a hinge and folded down, making the image temporarily disappear from the large end of the tube. This demonstrated the particle aspect of the radiated energy.

The Crookes tube by the turn of the century morphed into the CRT, whose defining feature was not, as frequently supposed, a hot cathode. Actually it was the addition of a phosphor layer at the large end of the tube which amounted to a screen.

Early CRT's were cold-cathode devices. Air was removed during manufacture, creating a 0.01 Pa to 133 nPa vacuum. (In a hot-cathode tube, the vacuum serves the dual purpose of preventing oxidation and immediate destruction of the filament and also permitting the unimpeded flow of electrons.)

Western Electric developed the hot-cathode CRT and began marketing it in 1922. In the hot-cathode tube, thermionic emission greatly enhanced electron flow from the cathode. This innovation laid the groundwork for radar as well as the modern oscilloscope.

In a hot-cathode tube, the cathode can be either directly or indirectly heated. In a directly heated cathode as in a fluorescent bulb, the filament *is* the cathode. But in vacuum tube amplifiers and CRT's to mention two examples, AC vibrations can be problematic because gain becomes a function of the utility wave cycle. The filament current can be rectified and filtered, but a simpler solution is to indirectly heat the cathode. In this very common solution, the cathode is a negatively biased metal cylinder that surrounds and is heated by the hot filament.

In an oscilloscope, the CRT brings it all together in the sense that signals and voltages at the tube's inputs form and characterize the trace that is seen by the user.

That being said, we should note that CRT's are nearly obsolete, used now only in a few niche applications such as displays in some military equipment and oscilloscopes for educational use. Flat-screen displays are more complex and difficult to understand, but they are less expensive to build, and they have other advantages. They are lighter, less bulky, consume less energy, and dissipate less heat. CRT's are more hazardous due to risk of implosion if mishandled, and there are problems in end-of-life disposal.

The CRT contributes to distributed capacitance, and since the deflection circuitry requires high voltages, which linger long after the equipment powers down, there is shock hazard at the anode. Electrolytic capacitors used for filtering rectified DC at the power supply output can maintain a lethal charge for hours if not days after power is disconnected.

Flat Screen

These and other factors have meant that flat-screen displays have displaced CRT's in television and oscilloscope technology. Liquid crystal display (LCD) and plasma display panels (PDP) were the two dominant flat screen varieties until around 2007, when LCD price reductions drove PDP from the market. Currently, oscilloscope flat screens are almost exclusively LCD. They work well, and PDP picture quality is at best minimally superior.

The term "liquid crystal" would seem a contradiction in terms. Here's how it works: At the rear of an LCD oscilloscope or TV is the backlight, originally fluorescent but currently consisting of an array of LED's, which provide a more uniform light source that is also more energy efficient, with less heat to dissipate. Situated in front (moving toward the viewer) is a pair of polarizer plates, with closely spaced lines, originally engraved but now created by chemical means. Light from the LED array is polarized by the rear plate, and since the plates are installed so that the lines are perpendicular to one another, the unit, without the liquid crystals, would be opaque, and the screen would appear dark. The liquid crystals are contained by two additional layers inside the polarizer plates, which align the two ends in such a way that they have a 90-degree twist, causing the polarized light in each individual cell to twist, negating the effect of the polarizing plates. (When the set is powered down, the screen still appears dark because the backlighting is off. When a powered-up LCD remains dark, look to the backlight and its power supply.)

Voltage applied to the crystals negates the effect of the alignment layers, causing the crystals to straighten out so that light is no longer conveyed, the screen going dark. All this happens individually at a pixel level, and that is what creates the display. There is no gun or vertical and horizontal displacement and scanning as in a CRT. Overall the flat screen, using LCD technology, compared to the CRT, is less expensive, and there is less shock and implosion hazard and less heat dissipation, so it is easy to see why flat screen has displaced CRT usage in oscilloscopes, computer monitors, and TVs.

Triggered Sweep

Howard Vollum (1913–1986), a major oscilloscope researcher who with Jack Murdock (1917–1971) and other colleagues founded Tektronix, was responsible for a major innovation known as triggered sweep. It changed the way oscilloscopes work and vastly increased their usefulness.

For single and nonrepetitive events such as the human voice, the display in a pre-World War II oscilloscope was a realistic representation. But prior to the invention of triggered sweep, an oscilloscope could not display in a coherent way a repetitive periodic signal such as a utility AC waveform. By convention, analogous to reading a book, an oscilloscope display begins at the left side of the screen and proceeds along the X-axis time base. If the screen was very wide and arranged to slide along in front of the viewer, it would be possible to observe a coherent trace of a repetitive periodic signal for some finite length of time. But constrained by the horizontal size of the display, depending upon the horizontal scaling, the instrument quickly, typically in a small fraction of a second, runs out of space at the right side of the screen. The oscilloscope deals with this situation by returning to the left side and beginning anew. Specialized circuitry blanks out the electron beam (here for simplicity we are talking CRT) so that the retrace will not be a distraction.

What is problematic is that when successive waveforms revert to the left side of the display, they rarely coincide, so a coherent trace does not appear. If you want to see what this looks like, in the triggering section of the front panel controls, increase the triggering level, indicated in the display by a straight horizontal line, until it is higher (in amplitude, along the Y-axis) than the peak voltage displayed. At this point, triggering is lost, and it will be seen that the waveform becomes unstable and incoherent. Triggering is also lost when a certain amount of noise, typically 30 percent, is added to a sine wave obtained from an internal arbitrary function generator.

Howard Vollum and his colleagues at Tektronix built circuitry into their Model 511 oscilloscope, introduced in 1947, that caused the oscilloscope to initiate a new trace always at the same point in the associated waveform. The user can set this triggering point. In a typical instrument, the triggering point can be situated on either the positive or the negative going slope. Also, the user can set the triggering level in volts.

The point right now is to elucidate the concept of triggered sweep. Later we'll examine in detail triggering concepts and see how the user can alter triggering behavior to extract relevant circuit information.

A further innovation vastly expanded oscilloscope functionality. In the mid-1980s, digital technology came to dominate the scene in oscilloscopes as in most of the world of electronics. Walter LeCroy, a gifted photographer and founder of LeCroy Corporation (now Teledyne LeCroy), built the first digital storage oscilloscope, drawing on experience gained in designing high-speed digitizing equipment for CERN in Switzerland.

The digital revolution in oscilloscopes changed the way signals are processed after initial attenuation or amplification at the analog channel inputs. The analog signal in this new technology is sampled at the ADC at a rate determined by an internal clock. This digital signal is placed in memory so that it can be displayed, measured, and analyzed even when the analog signal is no longer present at the inputs.

Concurrent with the flat-screen display, virtually all oscilloscope manufacturers have shifted to a digital storage architecture, greatly enhancing the usefulness of the instrument.

Fourier Transform and the Oscilloscope

The Fourier series, a prelude to Fourier analysis and the Fourier transform, very accurately describes the behavior of oscillating waves. Taken together, this ensemble of theories constitutes a powerful tool for viewing ways in which matter and energy interact in waves, expressed as functions, and displayed as waveforms.

All of this was laid out in Joseph Fourier's *Analytic Theory of Heat* (1822), in which he presented a radically new perspective for understanding the wave nature of periodic signals. Of course at the time, his work could not have been understood as applicable to electrical current. Instead, the focus was heat flow.

The oscilloscope for the most part works in either of two modes – time domain or frequency domain. (Some instruments do both simultaneously in split-screen format.) Virtually all modern oscilloscopes can cycle quickly from one mode to the other, showing the same signal, when the operator presses the appropriate button. It is in translating back and forth between time and frequency domains that the Fourier transform becomes operative, as we shall see.

In the time domain, a waveform corresponding to a signal is depicted as the graph of a moving point plotted against two axes that intersect at the center of the screen. The vertical or Y-axis is calibrated to show amplitude in terms of volts, fractions, or multiples thereof. The horizontal or X-axis is calibrated to show the passage of time in seconds or fractions or multiples thereof. The simultaneous realization of these two values is a point that may be in any of the four quadrants. And since time is elapsing and the amplitude of the signal is usually varying, the point is moving so as to trace a line that often displays as a periodic waveform. (If the signal is unvarying DC, the trace consists of a flat horizontal line.)

In the frequency domain, this same signal can be displayed with the X- and Y-axes defined and calibrated differently. In this mode, the oscilloscope display has a radically different appearance, even when the identical electrical signal is applied to the input. In the frequency domain, rather than the passage of time, the X-axis is defined and calibrated to show a range of frequencies.

If the signal at the oscilloscope is a sine wave, all the power is concentrated at one frequency, known as the fundamental.

In the time domain, the sine wave has a distinctive appearance. When the amplitude is least, the rate of change is greatest, and when the amplitude is greatest (either in a positive or negative direction, i.e., without respect to polarity), the rate of change is least. At the positive and negative peaks, the trace is theoretically horizontal, and as it crosses the X-axis when the instantaneous amplitude is zero, the trace is theoretically vertical. These horizontal and vertical segments are infinitely small, having no finite duration.

A sine-wave trace consists of a fine line that conforms to the signal. As the signal deviates from a sine wave, the trace acquires first small irregularities and then increasingly non-sine wave features such as clipping, square corners, irregular rise and fall times, etc. Even in a pure sine wave, there is slight irregularity, due to the noise floor of the oscilloscope. This noise floor is always present, even in the best instruments. It is a consequence of thermal activity in any device or conductor that has resistance, and it is cumulative, beginning in the equipment or natural process that generates the sine wave, increasing as the signal progresses through the oscilloscope input and processing circuits, thence to the display and eye of the beholder.

A case can be made that without noise or the finite size of the pixels in the display, the trace of a theoretically pure sine wave would not be visible because it would have infinitesimal width.

In the frequency domain, the noise floor is more prominent. It consists of a jagged line, roughly horizontal, and constantly fluctuating, running the width of the screen. In frequency domain display of a sine wave, since all the power is in the single fundamental, the display consists of a single slender spike that may be located anywhere along the X-axis, depending upon how the instrument is currently configured. The height of this spike above the X-axis depends upon the amplitude of the signal connected to the oscilloscope input and also upon the user-selected vertical gain.

Oscilloscope users as they gain knowledge and expertise in the use of this multifaceted instrument generally focus increasingly on the frequency domain mode because it makes available valuable information that cannot be readily discerned by gazing at a trace displayed in the time domain. Fourier has this to say:

A function of time, that is to say a signal, regardless of its complexity, can be expressed as the sum of a number of sine waves. The Fourier transform, making use of a highly complex mathematical process, allows us to decompose any function of time into a function of frequency. It is a two-way process. The function of frequency can be transformed into a function of time. This reversible transformation can be performed any number of times with no loss of information.

The mathematics was greatly simplified by the development of the fast Fourier transform (FFT), a set of algorithms that facilitates the analysis-synthesis two-way process, making it accessible to a college-level math student. The FFT rapidly computes Fourier transformations by factorizing the discrete Fourier transform (DFT) matrix into a product of factors that are mostly zero. The work was begun in the early nineteenth century, becoming generally available in 1965, just in time for the digital oscilloscope.

In today's test and measurement environment, we can perform FFT so that time- and frequency-domain representations are viewed in rapid succession. Oscilloscopes built by different manufacturers vary, but the basic ideas are substantially consistent. Time domain is the default. To move to frequency domain, the user can press the Math button and navigate to FFT. Pressing the associated soft key, the signal instantly displays in the frequency domain. A mixed-domain oscilloscope will display the signal in both domains simultaneously, in split-screen format. In Chap. 7, describing troubleshooting techniques, we'll see how this capability plays out to help the technician isolate defects that cause equipment malfunction.

Spectrum Analyzer Basics

Of the many test and measurement tools that are available from vendors, the oscilloscope is a versatile and powerful instrument, permitting users to view signals in time and frequency domains. A somewhat more specialized instrument, the spectrum analyzer, also has a place in every electronics laboratory and well-equipped shop. It is similar to the oscilloscope in that it is capable of displaying electrical signals, but it is primarily a frequency domain machine. While lacking the oscilloscope's versatility, the spectrum analyzer surpasses it in advanced features and specifications.

Using the spectrum analyzer involves a steeper learning curve than using the oscilloscope, but for one who is adept at operating an oscilloscope, a few hours at the controls will go a long way in solving that problem.

Additionally, the spectrum analyzer is substantially more expensive than an equivalent level oscilloscope. But there is a way out of this difficulty. The PC-based spectrum analyzer, as we shall see, is a low-cost alternative that does not skimp on features, specifications, and performance.

The spectrum analyzer has evolved over the years, successive versions appearing with vastly different block diagrams and capabilities. The first model, still widely used, is known as a swept-spectrum analyzer. The operating system deploys super heterodyne technology, an architecture invented during World War I and employed in most radio and TV since that time up to and including the present. It is an efficient and cost-effective method of processing a modulated RF signal from antenna to speaker terminals – a little complex in concept but simple in implementation. Here's how it works:

An AM or FM signal from the antenna is applied in the usual manner at the tuner input, where it may undergo one or more stages of RF amplification. This output is fed to a mixer, which consists of passive diodes or active transistors, which in addition to their intended mixing function also contribute gain.

Besides the modulated RF carrier signal, an oscillating electrical signal from a local oscillator is applied at the mixer input. When two waveforms are juxtaposed in this fashion, additive and difference signals appear, so at the output of the mixer, a total of four signals are present. The one we are interested in is the difference signal. It is a simple matter to separate it out and suppress the others by means of a band-pass filter.

This difference signal, now known as an intermediate frequency (IF) signal, has two characteristics which make it valuable in the superheterodyne radio receiver application. One of these is that it is lower in frequency than the original RF signal. This is good because the signal can be efficiently amplified without battling the twin demons of capacitive and inductive reactance.

The other useful aspect of that IF signal is that it is a single, stable frequency regardless of the portion of the RF spectrum that is selected for reception. This is accomplished by varying the oscillator frequency as different broadcast stations are tuned in. A component of an old-style, nonelectronic tuner is the two-gang variable capacitor ("condenser") consisting of two sets of closely spaced metal plates mounted on a single shaft so that the capacitances of both devices vary in concert. The oscillator frequency is always fixed relative to the RF, and accordingly IF never varies. This fixed frequency is advantageous because all post-mixer amplification stages process a signal that varies only in amplitude. Then, a detector stage demodulates the IF carrier, leaving an audio frequency (AF) signal to drive the speaker or video and synch circuits in a TV.

The swept-spectrum analyzer also has a local oscillator. It uses superheterodyne technology to down-convert one segment of the input spectrum at a specific rate by sweeping the oscillator. Because this device is voltage controlled, its frequency can be varied at an appropriate rate, which is strikingly apparent in the display.

Resolution bandwidth in this instrument derives from the bandwidth of the band-pass filter, and this establishes the instrument's bandwidth. Smaller bandwidth equates to greater spectral resolution. Update speed and frequency resolution are inversely related, and this determines whether or not the user can discriminate between closely spaced frequency components. The bottom line, in a swept-frequency analyzer, is that a rapid sweep rate will reduce displayed amplitude and displace the frequency that is displayed.

These limitation in the swept-spectrum analyzer motivated a shift to the FFT-based instrument. Instead of sweeping the frequency span, the FFT-based spectrum analyzer uses an ADC to perform digital sampling at a rate equal to or exceeding twice the frequency of the signal being investigated. That rate is necessary to comply with the Nyquist sampling theorem.

This theorem states that in order to capture all the information from a continuous-time signal of finite bandwidth, the sampling rate must be at least twice the frequency of the analog signal at the input. If the sampling rate is less, aliasing will occur, and information ambiguities will arise, invalidating the digital record.

Satisfying the Nyquist criterion guarantees that the digital version will be accurate. Under certain circumstances, when other conditions are met, varying degrees of fidelity can be achieved, which is why signal compression is feasible. But generally, the Nyquist criterion is applied in analog-to-digital conversion.

After the signal has undergone conversion, an FFT-based spectrum analyzer chooses a portion of the spectrum, but in this instrument, it is not swept. The FFT-based spectrum analyzer reduces the required sampling rate (because frequency is less) so that the instrument can process the entire range of signals. The FFT-based spectrum analyzer operates in this non-sweeping mode in order to avoid missing short-duration events, and in this it is largely successful.

But the real-time spectrum analyzer represents a significant step forward. For now, it is the best available spectrum analyzer. It displays both dynamic and transient RF signals, and that is precisely what is needed to assure full compliance in new electronic products is still in the prototype stage. The real-time instrument is capable of triggering on any desired RF signal, which can subsequently be placed in memory so as to be available for analysis in both time and frequency domains. That is because the real-time spectrum analyzer examines the entire spectrum with no gaps or missing pieces. It looks at the signal in an intact IF that equates to the instrument's full bandwidth. Accordingly, this instrument permits the user to capture and analyze brief transient events that are not available in swept-spectrum and FFT-based analyzers. The extremely fast processing engine can display the entire bandwidth without gaps, its ADC digitizing the entire bandwidth that falls within the passband. Moreover, the memory is capable of acquiring data throughout the measurement period.

In viewing the frequency domain display from a given source, the user can see spurious signals including harmonics, intermodulation products, and noise. By monitoring them, one can determine whether they comply with amplitude and frequency requirements. Bandwidth measurements of modulated signals will ascertain whether they occupy the space within a user-imposed mask.

After a signal under investigation has been applied at the input of a spectrum analyzer, the task is to obtain a coherent and meaningful frequency domain display that conveys the desired information. There are variations among spectrum analyzers built by different manufacturers, but the general idea is to begin by going into the settings menu and adjusting frequency and amplitude levels. (The settings menu is typically associated with a prominent gear icon.)

One approach is to begin with the center frequency. Using the numeric keypad, the center frequency of the signal under investigation can be entered into the appropriate field. This locates the signal at the center of the display. The main signal and harmonics appear, the height of the spikes corresponding to their amplitude in power as shown by the logarithmic decibel scale on the Y-axis.

Another approach is to begin by selecting the span. This operation determines size in terms of frequency of the spectrum at the input. The span is calibrated in hertz per division.

A third approach is to set the start and stop frequencies. When these values are entered into the appropriate fields, the span will automatically set itself. Always, the difference between start and stop frequencies is equal to the span.

In addition to setting the frequency parameters, the spectrum analyzer user has to adjust the instrument's gain in order to create a meaningful display. There are two gain controls, RF attenuator and IF gain. These controls must be set together and balanced in order to obtain a usable display. If the signal level at the mixer is excessively high, all succeeding stages may become overloaded. If gain is increased, then noise at the input is amplified, and noise levels on the display become higher. This background will mask out lower-level signals. For this reason, the amplitude levels need to be carefully coordinated.

Also, filter bandwidths determine the overall bandwidth of the instrument. In many spectrum analyzers, the IF filter is labeled as resolution bandwidth, and the associated adjustment sets the spectrum resolution.

Triggering in a real-time spectrum analyzer is not subject to delay, so there is continuous signal capture and both time and frequency domain analysis. The real-time spectrum analyzer implements an ADC that provides low noise and high dynamic range. All these features make for a hefty cost, far more than the price of a comparable oscilloscope.

Tektronix has largely mitigated this unfortunate situation by bringing out its RSA306 PC-based spectrum analyzer. The rationale is that most potential users possess or have access to a PC. The Tektronix module together with a signal source cables to the PC, which powers and interacts with the module through a standard USB 3.0 cable. Since the module has no external controls, moving parts, or internal electrical contacts, there is little chance of damage unless it is grossly overloaded. The ruggedized enclosure is suitable for outdoors or the factory floor.

An essential element in this assembly is the SignalVu-PC software, supplied with the PC-based spectrum analyzer in a flash drive or as a free download from the manufacturer's website, where a user manual and other documentation are also available even if you have not purchased the spectrum analyzer. The overall cost is far less than that of a bench-type instrument.

Operating a spectrum analyzer is somewhat less intuitive that operating an oscilloscope, but users find the rewards substantial, and for many types of signal analysis, it is essential.

In this opening chapter, the intent has been to provide an overview of oscilloscope technology without overwhelming the reader in details. While subsequent chapters will become increasingly technical, the aspiration is to progress in an orderly fashion so as to provide background information as needed in order to address specific applications in signal analysis.

References

Gibilisco, Stan, *Trigonometry Demystified*, Second Edition, McGraw-Hill, 2012
Hickman, Ian, *Oscilloscopes*, Fifth Edition, Newnes, 2001

Chapter 2
Semiconductors Inside the Oscilloscope and as Objects of Inquiry

Abstract Semiconductors inside the oscilloscope and as objects of inquiry. Why CMOS technology has endured. Oscilloscope block diagrams and functional details.

In future chapters we'll be talking about ways in which an engineer, technician, or student can use a digital storage oscilloscope to visualize current flow and waveforms in circuits, equipment, and networks. All of that is a rather vast subject (Hickman 2001).

Before covering that set of subtopics and as a necessary precondition for understanding them, in this chapter we'll look at the block diagram of a modern oscilloscope. We'll also undertake a discussion in some detail about how this instrument works. We'll consider how it can transform and manipulate electrical inputs as displayed on the screen.

Some Basics

To begin, let's review a few basics. The oscilloscope is a complex assembly of electrical circuits in turn composed of many components. It is fair to assume the reader understands how passive components (resistors, capacitors, inductors) work. But most of us can use a good review of active devices, particularly solid-state semiconductors (Gibilisco and Monk 2016). In that spirit, here's an overview:

"Semiconductor" implies a device that falls somewhere between a sheet of glass, which is close to a perfect insulator, and a copper bar, which is close to a perfect conductor. A resistor would qualify, but that is not what is meant by this specialized term. As used in an electronic context, a semiconductor is a device that conducts under certain electrical conditions and does not otherwise conduct.

The simplest semiconductor is a diode. A water analogy is often useful. A diode is like a check valve in a water pipe. Both permit flow in one direction but not in the other direction. A diode has two terminals, anode and cathode. When positive voltage is applied to the anode and negative voltage is applied to the cathode, the device is said to be forward-biased. Current flows through it. When these electrical

© Springer Nature Switzerland AG 2020

D. Herres, *Oscilloscopes: A Manual for Students, Engineers, and Scientists*,
https://doi.org/10.1007/978-3-030-53885-9_2

connections are reversed so that positive is connected to the cathode, the diode is said to be reverse biased, it has high resistance, and very little current flows through it.

When AC is applied to a diode, the device conducts only during the time it is forward-biased. Depending on the design of the diode, this selective behavior can oscillate at very high frequencies.

Because of its semiconducting property, the diode has a great many applications in electronic circuits:

- When placed in one of the conductors in a two-wire circuit, it constitutes a half-wave rectifier, converting AC to an intermittent pulsating DC. The half-wave rectifier is not much used due to its inefficient output and the difficulty in filtering to remove ripple. Instead, four diodes (six for three phase) are configured to make a full-wave rectifier.
- A blocking diode takes advantage of forward-reverse bias behavior to selectively connect-disconnect two portions of a circuit. A very interesting example is in a wind system where the generator is cabled to a battery bank. When a drop in wind speed causes RPM and output voltage to drop below a certain level, the DC feeds back through the generator windings, causing the armature to turn. (A DC generator can also act as a DC motor, and this is called motoring.) This behavior depletes the backup batteries as well as baffling passersby, who see the turbine spinning when there is no wind. The remedy is to install a blocking diode in series with the generator and battery bank so that current does not flow in the wrong direction. A single-pole button labeled "motor" can be added in parallel to and bypassing the blocking diode for the purpose of testing. If the generator will motor in no wind, the entire system is functional.
- As outlined in Chap. 2, a diode is the defining component in a superheterodyne mixer, where two frequencies are combined to beat against one another, producing sum and difference frequencies.

Silicon-Based Transistors

Now we'll consider the subjects on a subatomic level to see how semiconductors are formed and why they work. This will be a prelude to an overview of the many transistor variations. Semiconductors have been constructed of silicon (Si), germanium (Ge), and gallium arsenide (GaAs). We'll restrict this discussion to silicon, which is the most widely used semiconductor material at present. Like other semiconductors, silicon has relatively few free electrons because the atoms are closely packed in a crystal lattice. Free electrons, as needed for electrical conduction, can move only under limited conditions.

Electrical conduction in silicon can be greatly enhanced by adding minute amounts of certain other substances, a process known as "doping." The impurities, typically one atom per ten million atoms of the semiconductor, may be added simply by passing the dopant in the form of a gas over the semiconductor material under carefully controlled conditions.

To understand what is going on in a semiconductor, we have to consider the nature of a crystal lattice and also the whole matter of electron orbits, especially valence.

There are numerous lattice types, with complex names like rhombohedral and orthorhombic. This has a lot to do with interplanar spacing and lattice vectors. Suffice to say that when atoms of any substance including silicon are arranged in a crystal lattice, they are tightly bound with few free electrons.

Electrons revolve around an atom's nucleus in the manner of planets circling the sun in our solar system. The only difference is that the planetary orbits in our system lie in the same plane. Electron orbits in an atomic nucleus crisscross in a seemingly random fashion. For this reason, they are said to occupy shells. Each of the roughly 100 known elements has a different number of electrons orbiting the nucleus.

Each shell can accommodate only a specific number of electrons, except for the outermost or valence shell, where the number may vary anywhere between one and eight. The number of electrons in each shell excluding the valence shell is given by the algebraic expression $2n^2$ where n represents the shell's number, beginning with the innermost, for which n equals one, and working out. (An exception is hydrogen, which has only one electron.)

The first shell has two electrons. The second shell has 8 electrons, the third shell has 18 electrons, and so on, except for the valence shell. The number of electrons in the valence shell determines how the atoms interact with one another. What makes silicon interesting and accounts for the fact that it is a semiconducting material is that it has four electrons in its valence shell. For this reason, silicon atoms are able to arrange themselves in a highly stable crystal lattice, wherein a single silicon atom shares electrons with four neighboring silicon atoms, so that all five atoms have valence shells that are fully occupied.

Pure silicon in its crystalline form is a near-absolute insulator. A critical step in transforming it to a semiconductor is slicing the crystalline material to make thin wafers, which are then doped by adding very minute amounts of specific impurities. The type of doping material and amount applied affect the semiconductor properties of the final product.

The simplest semiconducting device (simple because unlike a transistor, it has only two layers with a junction where the layers are bonded together) is the diode. Electrical leads are affixed to the far ends of the layers, away from the junction. One layer is composed of P-type semiconducting material, and the other layer is composed of N-type semiconducting material. (P denotes positive, and N denotes negative.) The lead that is connected to the P layer is known as the anode, and the lead that is connected to the N layer is known as the cathode. In electronics, generally speaking, anode refers to a positive pole, and cathode refers to a negative pole, as in batteries and vacuum tubes.

The diode semiconducting behavior varies with the polarity of the voltage applied to anode and cathode, and it has everything to do with what is occurring in the P and N layers and in the junction on a subatomic level.

Semiconducting activity takes place because of the presence of charge carriers, which may be negative electrons or positive holes. When these charge carriers migrate to the junction, the device becomes conductive. What makes them migrate is the external voltage that is applied through the leads to the semiconducting layers. When forward-biased, this voltage repels the charge carriers. (Negative voltage repels electrons, and positive voltage repels holes.) The charge carriers are squeezed in toward the junction. When reverse biased, the charge carriers are pulled away from the junction, because the negative voltage applied to the anode and the positive voltage applied to the cathode attract, respectively, holes and electrons. Deprived of charge carriers, the junction and the semiconductor as a whole do not conduct.

Diode Parameters, Especially Inductive and Capacitive Reactance

All diodes function according to these basic principles. But the physical construction, amount of doping, and other construction details determine electrical behavior, parameters, and applications. First, there is the all-important matter of frequency, which directly affects capacitive and inductive reactance.

Any electrical device possesses in some measure both inductance and capacitance and, it follows, inductive and capacitive reactance. These properties are often unintended and harmful.

In a two-wire transmission line or electrical circuit that feeds into two nodes, unintended capacitance is generally a parallel phenomenon, and unintended inductance is generally a series phenomenon. Therefore, unintended low capacitive reactance is "bad" because it tends to shunt out the signal at high frequencies, and unintended high inductive reactance is also "bad" because it attenuates the signal at high frequencies.

In using a diode in a high-frequency circuit, parasitic capacitance enters the picture because the leads and terminals act like plates in a capacitor, causing it to act like a conductor even when reverse biased, and the leads and terminals exhibit inductive capacitance causing it to attenuate the circuit if the diode is in a series application.

Diodes that are intended to operate in a high-frequency circuit can be designed physically so as to minimize inductive and capacitive reactance.

Transistors, like diodes, are semiconducting devices, and they operate according to the same subatomic principles. The difference, however, is that (with exceptions) they have three semiconducting layers with three leads rather than two as in a diode, and they have two semiconducting junctions rather than one. All of this makes for more complex operation and many more parameters, but it also opens up a vast potential for diverse applications in amplification, oscillation, and switching.

The first widely used of these components were bipolar junction transistors (BJT's). Most BJT's have three layers, collector and emitter at the two ends with base situated between them. The geometry is all-important: the emitter and collector each have one junction opposite the lead. They share these junctions with the base, which therefore has half ownership of two junctions with one lead typically situated midway between the junctions.

In the NPN BJT, the two outer layers, with collector and emitter leads attached, are composed of N-type semiconducting material, and the middle layer, with base lead attached, is made of P-type semiconducting material.

In the PNP BJT, the two outer layers, with collector and emitter leads attached, are composed of P-type semiconducting material, and the middle layer, with base lead attached, is made from N-type semiconducting material. These two transistor types work essentially the same but with all polarities and directions of current flow reversed. The PNP devise is less used because holes, the majority charge carriers, are less efficient than electrons. Therefore, in the discussion that follows, we will focus on the NPN BJT.

A transistor can be modeled as three diodes connected in the configurations shown in Figs. 2.1 and 2.2. These configurations are known as Ebers-Moll models.

The BJT, with multiple conduction modes, conforms to the venerable and very useful Ebers-Moll model. NPN transistors are equivalent to two diodes whose anodes join. PNP transistors are equivalent to two diodes whose cathodes join. In both versions of the Ebers-Moll model, I_B, I_C, and I_E are, respectively, the base, collector, and emitter currents, and α_F and α_R are the forward and reverse common-base current gains.

Fig. 2.1 Ebers-Moll model for NPN transistor. (Wikipedia)

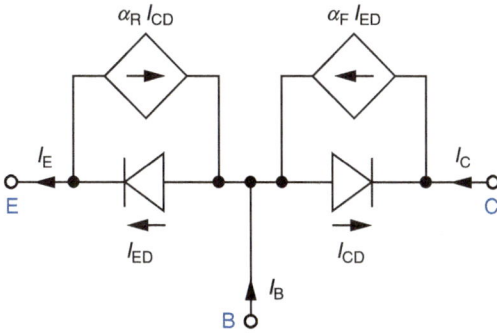

Fig. 2.2 Ebers-Moll model for PNP transistor. (Wikipedia)

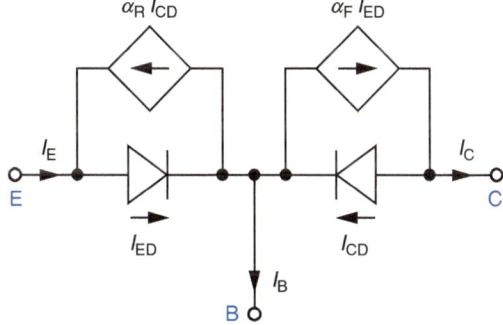

This diode model is accurate in terms of polarities and biases and is useful as a guide for preliminary diode tests using a multimeter in the test mode. However, it must be emphasized that you cannot build a transistor using diodes. The polarities will be correct, but the device will not amplify, oscillate, or function as a switch. The reason is that the semiconducting junctions are not actually shared, just superficially connected.

On a schematic, transistor pin notations conform to well-established engineering conventions (Gibilisco 2014). With these firmly in mind, reading circuit diagrams becomes far easier. At transistor terminals as depicted in a schematic, voltage with respect to ground is denoted by a single uppercase subscript. V_C, V_B, and V_E are voltages to ground respectively at the collector, base, and emitter.

Voltage drops between two terminals are denoted by V with a two-character subscript, all uppercase. For example V_{BE} is the voltage difference between base and emitter. But what if the two subscript characters are the same, as in V_{CC}? That denotes the positive power supply voltage at the collector, and V_{EE} is the negative power supply voltage at the emitter.

Other transistor schematic conventions are helpful in identifying the type of device and its pinouts. Referring to Fig. 2.3, the schematic diagram of an NPN transistor, the base connection is self-evident, a straight-line segment with a stylized lead perpendicular to its midpoint. The other two leads, collector and emitter, are also connected to the line segment, not parallel to one another, but diverging.

Referring to Fig. 2.4, the schematic diagram of a PNP transistor, the base, collector, and emitter are similar, the only difference being in the direction the arrowhead points.

Later we will see how the schematic for an FET differs, so that these two basic transistor families can be distinguished from one another even when the devices are not labeled in the schematic.

The emitter can always be identified by the fact that its lead has an arrowhead. If the arrow points out, the transistor is type NPN. If it points in, the transistor is type PNP.

The BJT, in addition to complying with the Ebers-Moll model, is a transconductance device. This means that the collector-emitter output current varies with the input voltage at the base. Small variations in base voltage cause large variations in collector-emitter current. That is the basis for amplification. In the Ebers-Moll

Fig. 2.3 Schematic of NPN transistor with emitter arrowhead pointing out. (Wikipedia)

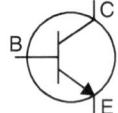

Fig. 2.4 Schematic of PNP transistor with emitter arrowhead pointing in. (Wikipedia)

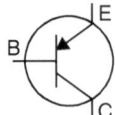

model, the output current is equal to the input current times the gain, which is represented by β (Greek beta). Rather than gain, however, when the BJT is considered as a transconductance device, the output current is proportional to the input voltage.

Field-Effect Transistors

For an understanding of FET's, the fundamental concept is that rather than current in the base circuit as in a BJT, there is an electrical field associated with the gate. The terminology differs between these semiconducting families, as follows:

- The FET gate is analogous to the BJT base.
- The FET source is analogous to the BJT collector.
- The FET drain is analogous to the BJT emitter.

These differences in terminology underscore the fact that what is happening on a subatomic level is altogether different in BJT and FET. The FET schematic, moreover, is easy to distinguish from that of the BJT. The gate, like the BJT's base, consists of a stylized lead that terminates at and is perpendicular to the midpoint of a short, heavy line segment, as shown in Fig. 2.5.

Enter the Ubiquitous MOSFET

That is where the similarity ends. The short line segment, in contrast to the BJT base, consists in a MOSFET of two parallel lines with a space between them, much like a capacitor, as shown in Fig. 2.6.

The gate and channel of a MOSFET do in fact have a thin insulating barrier between them, resembling the dielectric layer between the plates of a capacitor.

A JFET schematic is a little different. Rather than the space separating gate a channel as in a MOSFET, a single-line segment represents both elements. MOSFET

Fig. 2.5 N-channel
junction gate field-effect
transistor. (Wikipedia)

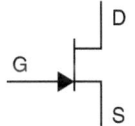

Fig. 2.6 Enhancement
mode N-channel MOSFET.
(Wikipedia)

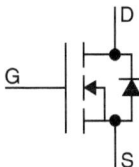

and JFET schematics both have a feature that makes them easy to distinguish from a BJT. It is that the source and drain in any FET come off from the gate line segment at 90-degree angles, whereas in a BJT, the collector and emitter form larger angles at the base, and they are not parallel. With these differences in mind, it is easy to recognize BJTs and FET's in schematics and to distinguish between JFET's and MOSFET's.

And just as the schematic symbols are unique and distinctive, the subatomic structures and electronic behaviors are unique for each of these devices, which make them useful in test instrumentation and in the circuits that are measured, displayed, and investigated using that instrumentation. This is particularly applicable for the oscilloscope as well as the spectrum analyzer.

FET's, both JFET's and MOSFET's, are for the most part three-terminal devices. Voltage, applied to the gate, creates an electrical field that determines the conductance of the channel, between drain and source. This electrical field, which oscillates in accord with the input signal, causes the channel conductivity to oscillate at the same (at times fluctuating, non-sinusoidal, and with abundant harmonics) frequency. Since a relatively strong bias from an external power source can be applied to the source-drain circuit, very significant gain is possible, as in a BJT device.

The operating principles in all these devices (BJT and FET – both JFET and MOSFET) are similar, but the numbers differ radically. Specifically, because the FET family of devices requires a minute electrical charge and close to zero current at the gate, the input impedance is *very* high. Because of this high input impedance, the FET is used in many applications where any loading of the preceding stage is not desired. A classic example is the front end of a multimeter in the volt mode. If the circuit under investigation has an appreciable impedance, a low impedance such as the electrician's familiar solenoid-based voltmeter (Wiggie) would not be suitable.

An FET may have either of two polarities. In an N-channel FET, conduction is by electrons, while in a P-channel FET, conduction is by holes. A hole is the absence of an electron, in effect a positive charge-carrying particle.

FET's can have either of two types of gates. There are junction FET's (JFET's) and metal-oxide semiconducting FET's (MOSFET's), which redefine the upper limits of internal impedance. Also, depletion-mode and enhancement-mode FET's are created by different types of channel doping. The N-channel enhancement mode MOSFET is the most suitable of these semiconductor varieties, and its use is widespread in today's electronic equipment. In most applications, the drain is positive with respect to the source. Current flows in the source-to-drain output circuit when the gate is positive with respect to the source. In this mode, amplification, oscillation, and switching functions are realized.

We have seen that BJT's are built as NPN and PNP devices, with analogous characteristics except for reversed polarities. Likewise, P-channel and N-channel MOSFET's are mirror images of one another, again with reversed polarities. However, P-channel MOSFET's like PNP BJT's are less used since the charge carriers, holes, have less mobility than electrons (Horowitz and Hill 2015).

The MOSFET's great functionality is due to the insulating glass barrier between gate and channel, which precludes flow of current. It is, however, transparent with respect to the electrical field present in the vicinity of the gate. (Glass, a near-perfect insulator, allows passage of another type of electromagnetic radiation – light.)

MOSFET's are vulnerable to any slight static charge that may be imposed on them in handling. For this reason, these sensitive devices are shipped from the manufacturer in antistatic packaging, often with the leads shunted. Technicians wear grounding antistatic bracelets when handling printed circuit boards containing MOSFET's.

Due to the very high gate-to-channel capacitance, the MOSFET exhibits an unusual behavior, which is that once turned on, the device remains in that state even after voltage at the input is removed. This is another consequence of the very high input impedance, often exceeding 10^{14} ohms. This interesting behavior is the basis for an easy way to check a MOSFET using a multimeter in the diode test mode. Begin by touching a grounded surface to remove any static charge you may have.

Then, connect the channel through source or drain terminal to the multimeter's negative lead and touch the positive lead to the gate. Next, shift the positive lead over to the source. A low reading will mean that the device is good at least to the extent that it is not shorted out.

Leaving the meter connected as above, simultaneously touch gate and source or gate and drain, which does the same thing due to the relatively low impedance between them. Now that the MOSFET is discharged, the meter should read high. This second part of the test indicates that the MOSFET is now nonconducting and that it is at least provisionally good. These ohmmeter tests are not totally definitive, but they rule out some common MOSFET failure modes.

For a true dynamic test, a laboratory-grade transistor tester is needed. This instrument applies signal and bias voltages at the inputs, and the tester analyzes the output and compares it to specifications.

You can determine the status of a transistor by means of a far less expensive instrument that can be fabricated in the shop. It exploits the properties of Lissajous patterns displayed in an oscilloscope in the XY triggered mode.

Lissajous Patterns

We have discussed two fundamental ways in which the oscilloscope displays one or more signals applied at the input(s) – time domain and frequency domain. There is another entirely different type of display, known as the XY mode (Hickman 2001). To see how this works in a triggered sweep digital storage oscilloscope, first connect the default sine wave from the built in AFG (or from an external source) to the Channel One input. The conventional time domain waveform is displayed. (The frequency domain waveform is displayed when you press Math and the soft key associated with FFT.)

Now press Acquire. The acquisition menu appears, and you will see that the XY display is currently off. Pressing the soft key brings up the XY display menu. Press the soft key associated with triggered XY, and you will see that XY display is currently off. Pressing the soft key brings up the XY display menu.

Press the soft key associated with triggered XY, turning it on, and you will see the Lissajous pattern for a single sine wave. It is a single flat horizontal line. Notice also that when triggered XY is activated, Channel Two is turned on in addition to Channel One. If the AFG signal is shifted to Channel Two, the single flat horizontal line becomes a single upright vertical line.

Where it gets interesting is when two separate, synchronized signals are applied to Channels One and Two. In a Lissajous display under these conditions, a signal is applied to Channel One. Rather than being triggered at a specified level in the rising or falling slope of its own waveform (what we might call being self-triggered, as in the time domain), it is triggered by a different albeit synchronized waveform that is applied to Channel Two. To emphasize, these two signals must be synchronized. They may differ in amplitude, frequency, and phase, but these parameters of the two signals must be related, and the signals must be synchronized if we are to see a stable Lissajous pattern.

Its worth looking at two 60 Hz utility waveforms of the same amplitude that are 180 degrees out of phase. One might be tempted to pick up the waveform from a branch circuit receptacle, connecting Channel One and Channel Two probe tips and ground return leads to opposite sides of the circuit. *Don't do this!* Such a hookup is hazardous and will cause a short circuit with high fault current. A full explanation appears in the Introduction to this book. This topic is so important that we'll revisit it here in case you skipped the Introduction.

The ground return lead of a probe connected to a grounded bench-type oscilloscope must never be allowed to contact a wire or terminal that is referenced to and floats above ground potential. That is because the ground lead when so energized becomes also connected to the oscilloscope chassis and through it to the premises equipment-grounding conductor and thence to the premises ground system. Such a connection will likely damage the equipment under investigation and the oscilloscope. It may also injure the user due to hazardous fault current and arc flash. For a detailed discussion of this important topic, please reread the Introduction to this book.

Where does this leave us? The object of this exercise is to apply two* 60-Hz signals that are 180 degrees out of phase (one is the inverse of the other) to Channels One and Two.

One method is to connect the probe tips and ground return leads for the two channels to opposite sides of the secondary winding of a 24-volt, class 2 transformer of the kind that powers the thermostat for a residential oil furnace. The primary and secondary windings of such a transformer are electrically isolated, which means that both sides of the secondary are isolated from ground. (Primary and secondary are magnetically, not electrically, coupled.) There are some cautions:

- Windings of an autotransformer are electrically coupled. This type of transformer carries grounding into the secondary.
- If additional loads are connected to the secondary winding and one or more of these loads is grounded, this will affect the status with respect to ground of the secondary.
- If the equipment under investigation has a metal chassis or enclosure that is in contact with a metal bench or conductive object that is grounded, that can cause a similar problem.
- If the transformer has a partial shunt between primary and secondary windings, grounding can invade the secondary. Check it out with an ohmmeter and assume nothing!

When appropriate cautions are observed, the two-channel probes can be cross-connected so as to obtain two 60-Hz signals of the same amplitude that are 180 degrees out of phase.

There is another method to obtain two out-of-phase signals of the same amplitude and frequency or for that matter of differing amplitudes and frequencies. This procedure consists of applying the sine wave, for example, from the AFG to Channel One, while the oscilloscope is still in the time domain. The sine wave appears in the display. Then press the Save/Recall Menu button. Set source as Channel One, using Multipurpose Knob a. Set the destination to R1, assuming that it is not occupied by a previously saved signal that you want to retain. Next, press OK Save. When the color changes to white (not one of the dedicated channel colors), you know that the waveform has been successfully saved.

Together with the original signal from the AFG, you can supply what is needed to make the desired Lissajous pattern. Use the horizontal shift knob to create any desired phase shift in a second sine wave. For example, simple Lissajous patterns can be created when the frequencies of the two signals are the same. In this case, and when the phase is not shifted, the pattern is a straight line sloping upward, left to right.

When the two signals are the same frequency with a phase difference of 90 degrees, the pattern is a circle. With a 180-degree shift, a straight line sloping down from left to right appears. When the frequencies are the same, the display consists of an ellipse or either of the two limiting cases, a straight line or a circle. Still more variations occur when frequencies and amplitudes are not the same, as shown in Figs. 2.7, 2.8, 2.9, and 2.10.

An oscilloscope in the XY triggered mode displays Lissajous patterns, and they can be used to determine the status of a transistor. These tests are performed using an octopus, which eliminates problems involved in testing components that are in a circuit board. To do an accurate multimeter test, all but one lead must be disconnected, and that involves soldering/desoldering operations, which can damage sensitive MOSFET's because of heat and circulating currents. (Even the best soldering techniques are problematic due to downsized components and board density.)

All this is very interesting, but does it have any practical significance, and is it relevant to our discussion of transistor parameters? Absolutely.

Figs. 2.7, 2.8, 2.9, and 2.10 Lissajous figures produced by two signals applied to Channels One and Two with the oscilloscope in XY mode. (Wikipedia)

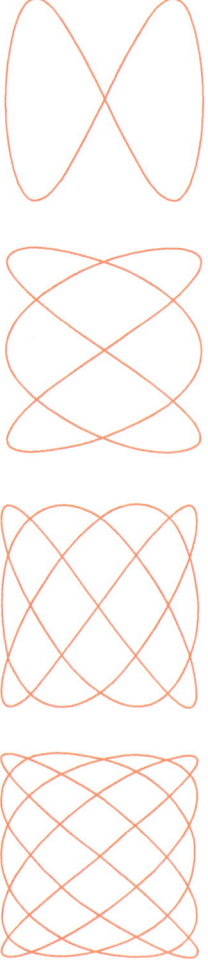

To display a Lissajous pattern, an oscilloscope must first be placed in the XY mode. Then, using specific hookups, transistors can be tested, and by interpreting the Lissajous pattern, the user can determine whether a transistor is good, bad, or questionable. The octopus, an aptly named mediating apparatus connected to the oscilloscope and semiconductor, is capable of testing a component that is soldered into a circuit board. Ohmmeter tests are useful, but they involve cutting leads and later resoldering them. Small components and densely populated boards are not user-friendly, and when there are lots of sensitive MOSFET's, the problem is compounded.

You can build an octopus from readily available and inexpensive parts. A 6.3-volt filament transformer and 3 resistors comprise this useful test instrument, shown in Fig. 2.11:

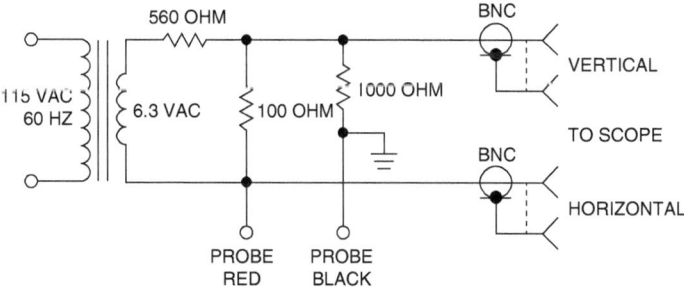

Fig. 2.11 The octopus, shown in the schematic diagram, connects to many types of electronic components including transistors, diodes, and integrated circuits. Its outputs go to two channels of an oscilloscope configured in the XY triggered mode. (Wikipedia)

Fig. 2.12 If the junction is good, the display will consist of a right angle. A 90-degree bend indicates optimal performance. Larger angles equate to questionable junction performance. 120 degrees indicate that the condition of the component is marginal, and if the angle is 150 degrees, the component will not function. (Wikipedia)

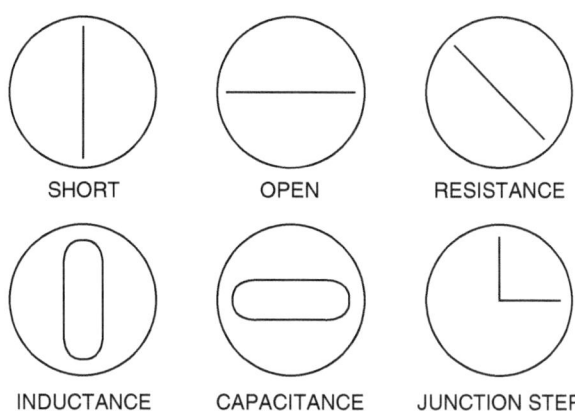

The octopus is easy to build. The 6.3-volt filament transformer can be taken from retired tube type equipment. The other components are inexpensive, and a metal enclosure with switch, pilot light, and AC cord, plus a dedicated analog oscilloscope with vertical and horizontal inputs, would be the basis for a valued transistor tester.

To evaluate the condition of a component, connect the octopus vertical output to the oscilloscope vertical input and connect the horizontal output of the octopus to the oscilloscope horizontal input. Connect the instrument's black lead to the PCB or chassis ground if applicable. When the red and black leads of the octopus are connected to junction terminals of the component under test, Lissajous patterns indicate its operational status. Here are some examples (Fig. 2.12):

To become adept at Lissajous interpretation, a good plan is to examine known good and bad components. Noisy potentiometers and questionable electrolytic capacitors produce distinctive Lissajous displays. Network integrity, variable frequency drives, and a wide variety of equipment types can be examined to good effect as well.

Anatomy of an Oscilloscope

In this section, we'll examine the inner workings of the triggered-sweep digital storage oscilloscope. We'll be looking at a block diagram and following the signal flow from analog channel inputs to flat screen display and also trace the power flow (Horowitz and Hill 2015). But first, for perspective, we'll consider an earlier two-channel analog oscilloscope with CRT display. Figure 2.13 shows the block diagram and typical front panel controls:

Notice that there are two analog channel inputs. These are typically BNC sockets located at the bottom of the center (vertical) section of the front panel. These inputs

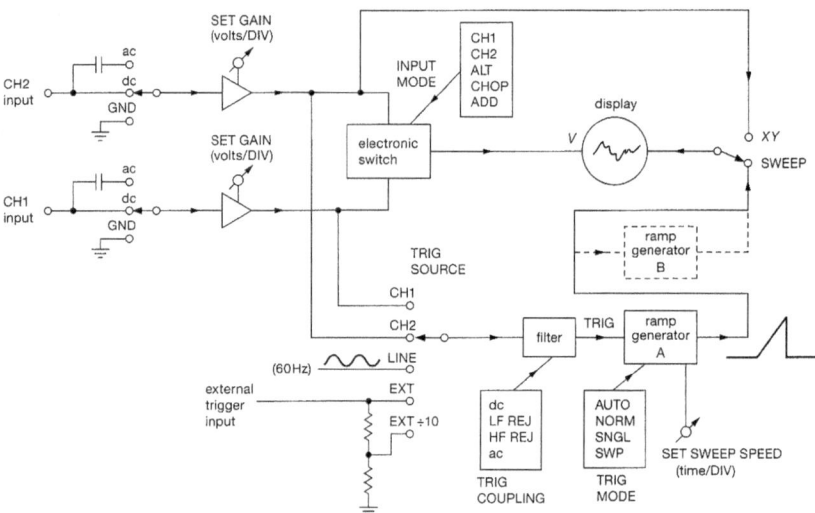

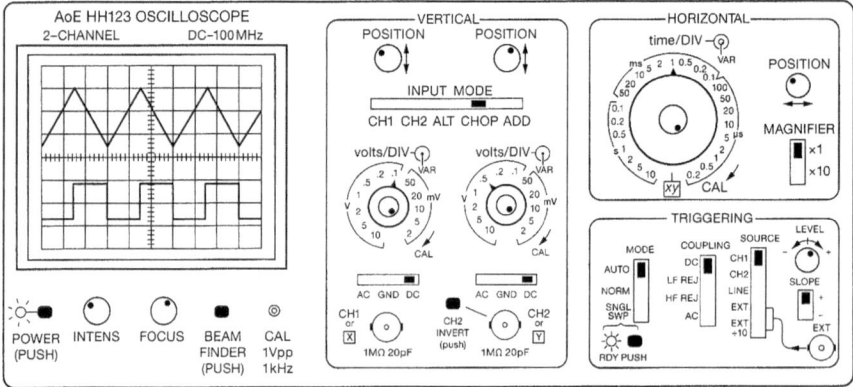

Fig. 2.13 Block diagram and front panel of an analog oscilloscope. (Tektronix)

Fig. 2.14 BNC cable
connector. (Judith
Howcroft)

can accept BNC cable connectors as well as probe cable connectors. A BNC (Bayonet
Neil-Concelman) connector, shown in Fig. 2.14, is used with standard coaxial cable.

This type of connector is reliable and easy to use. A quick quarter turn secures
the connection, as opposed to a conventional coaxial connector, whose threads are
more difficult to start, and they take forever to fully engage. BNC connectors meet
50 and 75 ohm characteristic impedance requirements (in two different noncompat-
ible sizes), and they are good for frequencies up to 4 GHz and voltages to 500 volts.

When the BNC cable is attached to the analog input channel connector, its outer
shell is electrically connected to the oscilloscope chassis and thence through the
power cord to the grounding terminal bar in the electrical service and ultimately to the
electrical system ground. The fact that this metal object is solidly grounded should be
kept in mind at all times when working with an AC-powered bench-type oscilloscope.
As mentioned elsewhere, if the ground return lead of a probe is connected to a wire
or terminal that is referenced to but floating above ground potential, sparks will fly.

Both analog channel inputs proceed separately to identical single-pole, triple-
throw switches. They permit the user to choose, individually, for each channel,
between AC coupling and DC coupling. Additionally, one or both of the signal lines
can be connected to ground, where there is no signal at the input, or we could say
that the signal is 0 volt DC. In any event, the corresponding display is a flat horizontal line that coincides with the X-axis. What is the purpose of this ground connection? There is nothing that corresponds to it in modern digital storage oscilloscopes,
but earlier developers must have had something in mind.

For one thing, in ground position the oscilloscope displays the 0 volt level. Also,
with signal disconnected from the instrument, the user can ascertain whether some
spurious or noise signal is coming into the oscilloscope through either of the inputs
or, alternatively, originating in the instrument's internal circuitry.

AC And DC Coupling

As for AC versus DC coupling, these alternate display modes profoundly affect
the appearance of the signal on the screen. As can be seen in the block diagram of
the analog oscilloscope, when AC coupling is chosen by means of either of the

three-position switches at the two-channel inputs, a capacitor is placed in series in the signal flow. Because of greater capacitive reactance at higher frequencies, in the AC coupling mode, any DC component of a signal is eliminated. This causes the trace associated with such a signal to drop down or rise up if the DC component is negative, so that it is situated at the X-axis notwithstanding a DC content at the oscilloscope input. This is helpful in visualizing a weak AC signal such as light ripple that is added to a strong DC voltage, precisely the situation when evaluating for purity the output of a power supply.

That being said, DC coupling provides a more realistic view of the actual signal at the channel input, so the switch should be ordinarily kept in the DC coupling position.

At the Channel Two input shown in the vertical section, an invert button can be seen. Also, in the schematic, there is an input mode switch, which permits the user to choose among Channel 1, Channel 2, Alternate, Chopped, and Add. In a two-channel machine, the ability to invert the signal is required in only one channel. If you push Add, you see the sum of the two signals. If you push Add in conjunction with Invert, you see the difference between the two signals. This is a key to understanding the differential amplifier and consequently the highly regarded and widely used op-amp.

All this is quite obvious, especially for users who have some experience with the digital storage oscilloscope in Add mode. Less self-evident, perhaps, are Alternate and Chopped. These modes are applicable whether or not a channel is inverted. In Alternate mode, the separate signals toggle in successive sweeps, whereas in Chopped mode, the display switches between the two, varying from once every 10 seconds to a million times per second. It is better to keep the oscilloscope in Alternate mode unless required to view a slow signal.

From the input mode switch, the Channel One/Channel Two signal proceeds to the display, ordinarily in an analog oscilloscope consisting of an old-world CRT. Within this clunky but reliable heavy glass envelope, the signal(s) applied at the Channel One and Channel Two analog inputs are applied to the vertical deflection plates, causing the bright spot on the screen to travel up and down, parallel to the Y-axis, in response to variations in the signal amplitude (Hickman 2001).

A double-throw switch enables the user to choose between XY, derived from the Channel Two signal line, and sweep derived from the ramp generator. The horizontal section assembly of stages is more complex than the vertical section, which simply relates the spot on the screen to the amplitude of the signal(s) present at Channel One and Channel Two analog inputs.

The horizontal section begins at a five-position switch that selects as trigger source Channel One, Channel Two, the AC line, or an external trigger input or by means of a voltage divider consisting of two resistors, the external trigger input divided by ten. The selected signal moves through a filter to the ramp wave generator. The filter includes a means for selecting the type of trigger coupling, which can be DC, high pass, low pass, or AC.

The ramp generator, as its name implies, generates a ramp wave, shown in Fig. 2.15, modified as described in the caption. It is applied to the horizontal deflection plates so as to create a time base for the instrument.

Fig. 2.15 This ramp wave, with the falling slope modified to make a vertical line that returns the trace instantly to the left side of the screen, is applied to the horizontal deflection plates. (Author's screenshot)

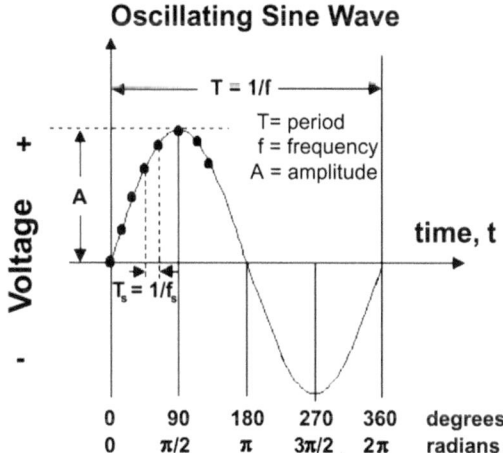

Regardless of its frequency, the ramp wave consists of a relatively slow rising voltage that, attaining a specific level, abruptly drops to zero. Applied to the horizontal deflection plates, this signal causes the spot on the screen to travel from left to right at a uniform speed. Reaching the right side of the display, the spot instantly reverts to the start of the time base and starts another trace.

On the front panel of an analog oscilloscope, the horizontal section has a subsection that is labeled triggering. Modes are auto, normal, and single sweep. Triggering is applicable for a repetitive or periodic signal. In this subsection, other than mode, we encounter coupling and source controls. The associated choices are shown in the block diagram. Also, level and slope at which triggering occurs can be adjusted. If the triggering level is raised to a level that is higher than the amplitude as shown on screen, triggering is lost, and the waveform breaks up to form multiple images from which usable information is difficult to extract.

Analog as well as digital oscilloscopes include trigger holdoff and bandwidth-limiting controls. Trigger holdoff disables triggering for a specified amount of time following a triggering event. The purpose is to prevent an irrelevant blip in a complex waveform from initiating unwanted triggering that occurs before the "real" event.

Bandwidth limiting, described in detail elsewhere, is a very effective method of eliminating noise that would otherwise obscure a signal of interest. It works because noise is generally a broad-spectrum phenomenon.

Triggered-Sweep Digital Storage Oscilloscope

In its time, especially after the introduction of triggered sweep, the analog oscilloscope was a marvel of electronic engineering. It successfully rendered clear images of electrical waveforms, not limited to fundamental voltage and current measurements like a multimeter.

The DSO finally prevailed around 1980. It quickly eclipsed its venerable analog ancestor, due to palpable advantages that will become clear as we examine the block diagrams shown in Figs. 2.16 and 2.17 of this amazing instrument.

At the input, as in the analog oscilloscope, the user is offered the choice of AC or DC coupling. Rather than mechanical switches with electrical contacts operated by knobs on the front panel, in the DSO the user makes this choice through a system of menus and contextual buttons. Typically, with a signal displayed, press the analog channel button. A menu appears below the display, and one item is Coupling. The soft key, so-called because it controls a variety of functions as different menus are invoked, toggles between AC and DC coupling. DC is the default and should be used most of the time. This is the way a DSO works. Control is electronic, accomplished through a system of menus, as opposed to electromechanical as in the analog oscilloscope. In the DSO, the only mechanical control is the power switch.

Next in the signal path, identical in both channels, are attenuators. Their purpose is to reduce the signal voltage by some consistent factor without affecting the frequency or the impedance at the analog channel inputs. They consist of a network of resistors and capacitors and in some designs an active component. The purpose of the attenuators is to protect the downstream circuitry. Of course, in measuring the amplitude of a signal, the amount of attenuation here as in the probe has to be factored in.

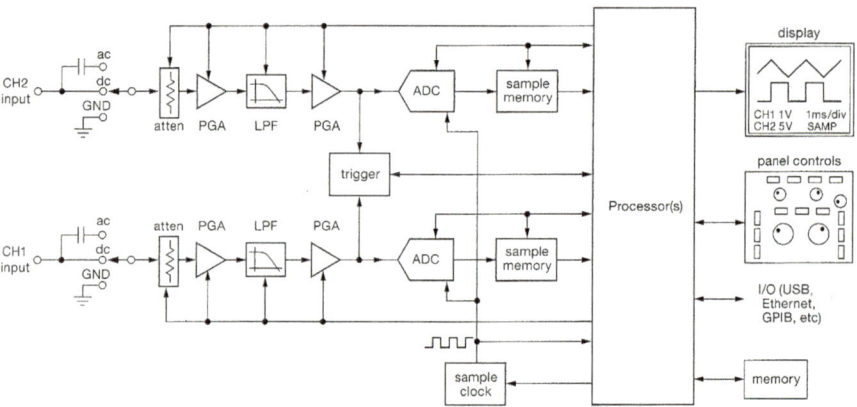

Fig. 2.16 This block diagram of a modern DSO, for simplicity, depicts a two-channel model. (Tektronix)

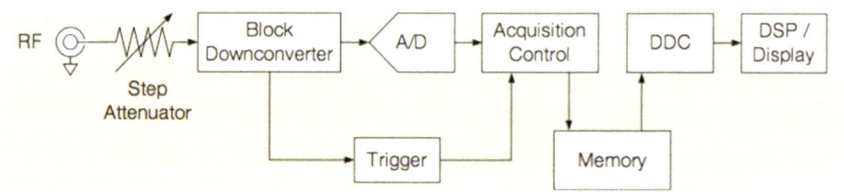

Fig. 2.17 Simplified block diagram of the Tektronix MDO4000 Oscilloscope. (Tektronix)

Following the attenuators are programmable gain amplifiers (PGAs), with two inputs, one from each attenuator and the other a control signal from the processors.

Attenuators and PGAs are designed so as not to make a bandwidth bottleneck in the signal path, which would happen with the introduction of unintended capacitance. These devices must not compromise higher-bandwidth instruments, and for that reason, they contribute to the higher cost of a high-bandwidth machine.

The next stage in preconditioning the signals, individually, in each channel, is the low-pass filter (LPF). Its purpose is to reduce the bandwidth of the signal (not of the overall oscilloscope) in order to eliminate noise and obtain a clear waveform. Noise is a broad-spectrum phenomenon, as can be seen in a frequency-domain display of a noisy signal, so bandwidth reduction works well to eliminate noise. To see it in action, access an AFG sine wave. In output settings, add 30 percent noise. Notice that the trace becomes indistinct and triggering is lost. Now press the appropriate channel button to open a new menu. Bandwidth is currently set at full, the default. Pressing the associated soft key opens the bandwidth menu. Depending on the bandwidth of the oscilloscope, bandwidth can be limited to various levels. An intermediate level will usually clear up the waveform only to a certain extent, while the lowest level will eliminate a substantial amount of noise and restore triggering. We'll have more to say on this topic in our discussion of noise.

This stage has two inputs, one containing the attenuated and PGA-modified signal from the channel input and the other input from the processor, which tells the low-pass filter the amount that the bandwidth should be limited.

The output from the low-pass filter is fed to another PGA, which further prepares the signal for digitalization. The output is fed to the trigger. The thing to be aware of here is that the oscilloscope triggers on the analog signal, prior to digitalization.

The second PGA output goes to the all-important analog-to-digital converter (ADC), which is a critical component in the DSO signal path. The schematic symbol is shown in Fig. 2.18.

The ADC samples the analog signal at a specified rate and creates a digital output, which is essentially a faithful version of the original signal at the analog channel input. Each channel has its own ADC, and they all (generally two or four) work independently and simultaneously to create digital signals that are conveyed to the processor.

The ADC samples the analog signal at a specific, uniform rate and assembles these samples to construct the digital signal. Before considering the process in a little more detail and examining the way in which it fits into the overall architecture of the oscilloscope, we'll look at the requirements and metrics of sampling in general.

Harry Nyquist, shown in Fig. 2.19, was born in Sweden but spent most of his working life in the United States. Long before Walter LeCroy built the world's first digital oscilloscope, Nyquist developed theoretical insights that eventually made analog-to-digital conversion possible.

Fig. 2.18 In an ADC, analog at the input becomes digital at the output. (Wikipedia)

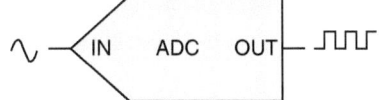

Fig. 2.19 Harry Nyquist (1889–1976). (Wikipedia)

While working at Bell Laboratories, Nyquist published papers on thermal noise, stability in feedback amplifiers, telegraphy, television, and related areas. What interests us in connection with an ADC had to do with information transmission through telegraph channels, which became the fundamental underpinning of information theory. Nyquist determined that the number of independent pulses that could be put through a telegraph channel in a given time interval could not exceed twice the bandwidth of the channel. This important insight, contained in *Certain Factors Affecting Telegraph Speed* (1924), evolved into what is known today as the Nyquist-Shannon Sampling Theorem.

This theorem addresses the process of translating an analog signal into its digital equivalent. It is accomplished by sampling the analog signal at uniform intervals. (Nonuniform sampling rates are sometimes valid provided the average rate is maintained.) It seems self-evident that a faster sampling rate (more samples in a given time period) will result in a more accurate digital signal. The Nyquist-Shannon sampling theorem quantifies the minimum sampling rate that is required for this process to occur with no loss of information. The theorem is applicable only when the original analog function has a Fourier transform that is zero outside a finite range of frequencies. That means simply that the bandwidth of the signal is not infinite. The theorem states that such a signal can be digitalized with perfect fidelity (no loss of information) when the sampling rate is twice the highest frequency component in the analog signal.

If digitalization is attempted at a sampling rate that is lower than required by Nyquist-Shannon, an imperfect digital record will result. Typically, the digital signal will exhibit aliasing.

The technical definition of aliasing as used in signal processing theory is that it is an effect that causes different signals to become undistinguishable when under sampled. Spatial aliasing can appear as irrelevant and distracting moiré patterns in digital images. In audio, under sampling can produce distortion or complete loss of the signal. In both cases, two remedies are available: Either the sampling rate can be increased or a low-pass anti-aliasing filter can be inserted in the signal path prior to sampling.

Returning to the digital oscilloscope block diagram, notice that the ADC has three inputs and one output. The inputs are the analog signal from the oscilloscope channel port, a control signal from the processor, and a timing signal from the sample clock, which sets the sampling rate.

To be appreciated, the ADC has to be seen in the context of post-digitalization processing. Figure 2.20 depicts a block diagram detailing stages that come after the ADC.

As you can see in Fig. 2.21, the sample rate is inversely proportional to the sample interval.

Sample rate and total time of a single waveform determine the amount of memory needed to store the signal in question.

An industry standard for most oscilloscopes, analog or digital, is ten horizontal divisions in the display, so the time duration equals the horizontal time scale multiplied by ten.

As for the ADC, it usually takes the form of an integrated circuit. ADC's are characterized in terms of bandwidth and signal-to-noise ratio. The bandwidth depends upon the sampling rate. It is important for the dynamic range of the ADC to exceed that of the signal at the oscilloscope input.

Resolution of an ADC is a key parameter. It corresponds to the number of discrete digital values that the device can output relative to the complete analog range.

Resolution is measured in volts or in number of bits. In the digital language, the number of bits corresponds to levels, expressible as powers of two. Thus, 2^8 bits equal 256 levels. In the analog language, as volts, resolution, known as span, equals voltage divided by the number of intervals.

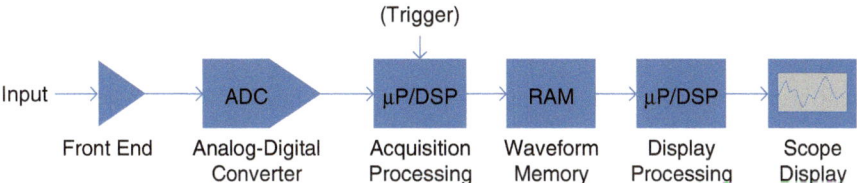

Fig. 2.20 The digital signal undergoes additional processing prior to display. (Tektronix)

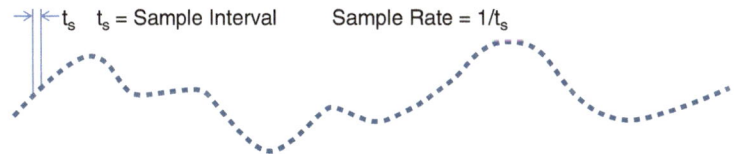

Fig. 2.21 The sample rate is dependent upon clock speed and sample interval. (Tektronix)

We described the three ADC inputs. The single output, as may be expected, goes straight into the sample memory, which, in turn, is conveyed to the processor. Each channel has a dedicated sample memory, connected to the processor. This is a one-way connection, as opposed to the single-system memory, with inputs from and outputs to the processor.

Digital signal processing takes place in the processor. This block has far more inputs and outputs than any other part of the modern DSO. In the block diagram, you can see that the processor directly communicates with the ADC's (one for each channel), the sample memory, trigger, sample clock, display, panel controls, I/O, and memory. All this is separate from front-end elements that precondition the signal prior to digitalization. A very important function the processor performs is compensating for individual channel differences that arise because of inevitable manufacturing differences. Below the maximum bandwidth rating, small variations can impact performance when individual channels are not in 100 percent agreement. Digital signal processing filters are a viable answer. The aspiration is to ensure inter-channel alignment at all levels of gain.

Variation in propagation speed at different frequencies is known as group delay. Since a high-bandwidth signal consists of an array of frequencies, group delay is certain to give rise to distortion. The processor measures and adjusts group delay in order to minimize any such distortion.

The processor furthermore enables bandwidth limiting, which reduces noise provided that the signal (if it has a high frequency) is not itself compromised. The processor also makes possible bandwidth boosting by increasing the gain in the front-end filter. Downsides, however, can include degraded signal fidelity because of increased noise, an effect that is the obverse of bandwidth limiting.

Front Panel

The Tektronix MDO3104 front panel, shown in Fig. 2.22, consists of a large display with a horizontal row and a vertical column of soft keys, so-called because their functions vary in response to menu choices invoked by the user. You can see this in action – just hit AFG, and the default sine wave will be accompanied by a wealth of soft key options, as shown in Fig. 2.23. Waveform, waveform settings, or output settings can be chosen in the bottom row. When one of these is selected, the appropriate vertical menu opens, and items can be chosen using the associated soft keys.

That's how it works. One of the keys to successful oscilloscope operation is for the user to become adept at navigating through these options.

Multipurpose Knobs a and (less-used) b change menu parameters. Numerical options can more quickly be set using the keypad wherever possible.

It is not recommended that a user open a digital storage oscilloscope enclosure to make alterations or repairs. Even advanced electronics technicians send a digital oscilloscope back to the manufacturer for servicing, which is surprisingly affordable.

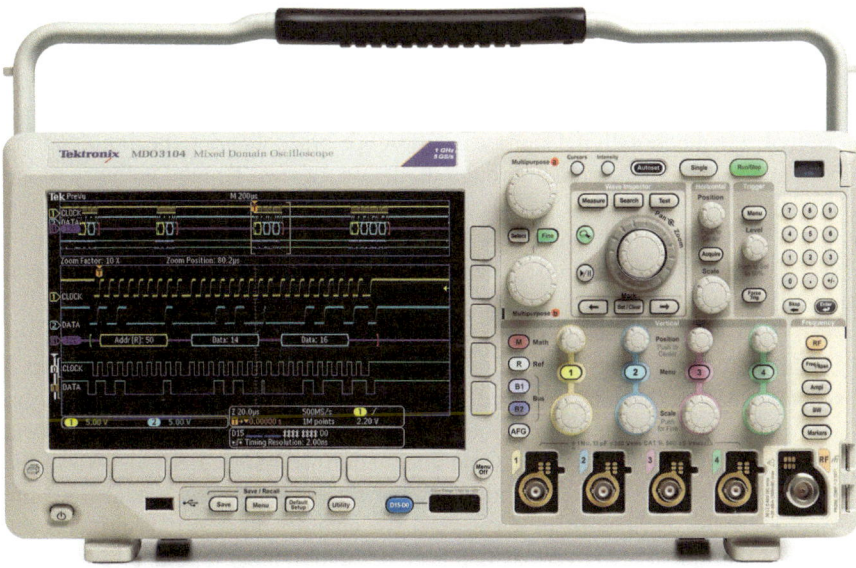

Fig. 2.22 Tektronix MDO3104 Oscilloscope front panel display and controls. (Tektronix)

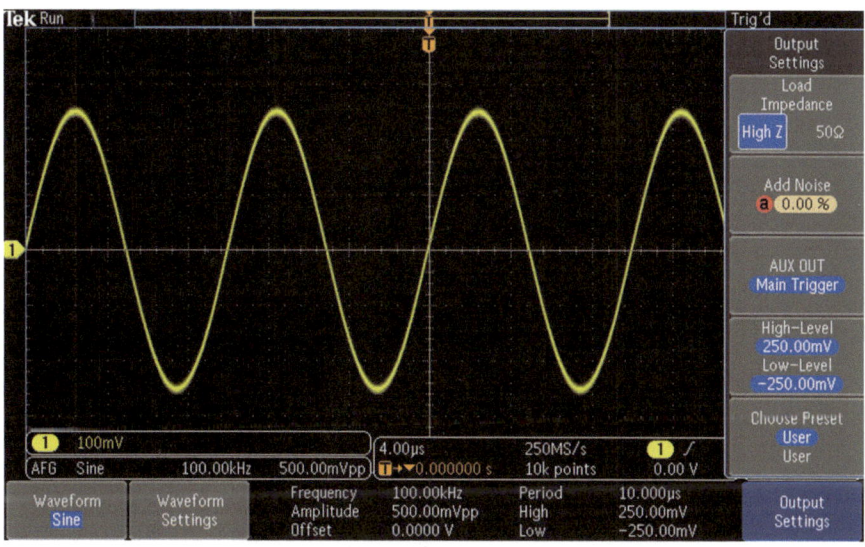

Fig. 2.23 AFG screen has vertical and horizontal menus. (Author's screenshot)

When the enclosure is opened, recalibration is required if you want to make accurate measurements, and it is not really feasible to solder components in the circuit boards as you might in a TV or stereo, due to the sensitive electronic environment where signals close to the noise floor are measured and displayed.

Nevertheless, to fully appreciate and competently operate an oscilloscope, a detailed knowledge of the inner workings is a great asset. Previously, in this volume, we have covered some basic material such as probe compensation, analog sampling in the ADC, and flat screen as an advance over CRT technology. Now we'll take a close look at a modern digital storage oscilloscope block diagram, beginning at the probes and following the signal through digitation, processing, and memory to the final display.

The signal from a device under test is applied to the channel input by means of a BNC cable or oscilloscope probe. The user must decide which probe to use, passive, active, differential, or current. Figure 2.24 shows a typical probe block diagram.

Out of the box, the oscilloscope is usually furnished with passive probes, one for each channel. As the name implies, these probes contain only passive components – resistors, capacitors, conductors, and the like. A passive probe can be adjusted by means of a slide switch on the probe body or from within the oscilloscope for a range of attenuations, primarily 1:1 (no attenuation) and 10:1. For most work, the 10:1 probe is suitable. It has a wide dynamic range. A 300-volt signal from the device under test is reduced to 30 volts at the channel input, which protects the oscilloscope from voltage overload and provides a high input resistance as seen by the circuit being measured, so that at moderate frequencies, the circuit is not loaded, i.e., the oscilloscope is invisible. Figure 2.25 shows a typical probe schematic diagram (Gibilisco 2014).

We have discussed the rationale and methods for passive probe compensation. Once this has been done, the operation does not have to be repeated. The oscilloscope knows the amount of probe attenuation. When a 300-volt signal in a device under test is probed at 10:1 attenuation, all oscilloscope measurement and waveform displays will show the 300-volt measurement.

The passive probe is good for general-purpose signal tracing. But depending upon the amount of accuracy that is required for a given application, at a signal regime in excess of 600 MHz, you'll have to start thinking about the far more expensive active probe. At these high frequencies, the low parallel capacitive reactance

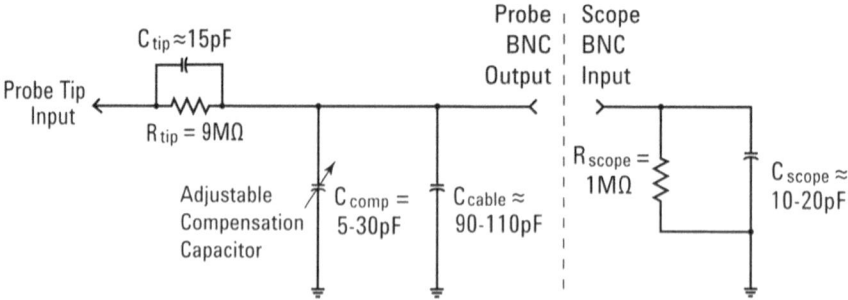

Fig. 2.24 Probe block diagram. (Tektronix)

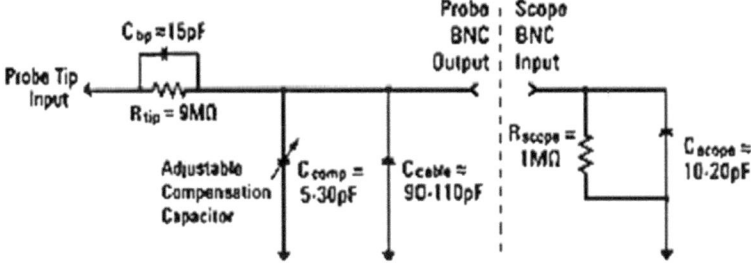

Fig. 2.25 Probe schematic. (Tektronix)

and high series inductive reactance of a passive probe negatively impact accurate measurements. The active probe, which requires external power to bias the semiconductors, is the way to go.

Passive Probes

By far the most used oscilloscope probe is the passive 10:1 attenuation probe (Horowitz and Hill 2015). It is appropriate when the frequency of the signal under investigation is less than 600 MHz. Impedance matching from probe tip to channel input port is critical. A mismatch gives rise to reflections, collisions, and loss of data. At DC to the mid-megahertz range, impedance consists primarily of resistance, but as frequency increases, capacitive and inductive reactance play an increasingly prominent role in signal attenuation.

The 10:1 passive probe offers less bandwidth and imposes heavier capacitive loading than an active probe. Probing the device does not load low-frequency signals being measured, and they are not significantly impacted from circuit board to channel input.

A further consideration in measuring a high-frequency signal is inductive reactance contributed by the ground return lead, which is proportional to its length. Short ground return leads should be used where possible when measuring higher-frequency signals.

Inductive loading, heavier in the passive probe, can be partially mitigated by using a short ground return lead and keeping it as straight as possible. (Even a partial turn at high frequency is in effect a fractional-turn coil.) By the same token, an active probe can tolerate a longer ground return lead where in some situations it may be necessary. Passive probes in general are satisfactory for debugging and troubleshooting in many circuits, but at 600 MHz plus, it may be necessary, depending on the amount of accuracy demanded by the application, to bring out the far more expensive active probe.

Active Probes

Active probes, as shown in Fig. 2.23, succeed in practically eliminating reactive loading of the circuit under investigation by means of a small solid-state amplifier contained in the probe body adjacent to the tip. The device has very high input impedance due to insulated gate technology. Drawing virtually no current, it has little discernable effect on the circuit that is probed.

Active probes are more costly. The Keysight 1 GHz N2795A active probe shown in Fig. 2.26 is $1111. Its input impedance is 1 megohm and 1 pF, and the dynamic range is 0 to ±8 volts. The Keysight 2 GHz N2796A active probe is $2253. Its input impedance is 1 megohm and 1 pF, and the dynamic range is 0 to ±8 volts.

From a practical standpoint, a major difference between active and passive probes is that active probes require power, either an internal battery or an external DC power supply.

In the active probe body, there are additional components. Besides amplification, there is signal filtering and also the capability for varying amounts of automation between probe and oscilloscope. When you plug the active probe cable connector into the analog channel input port, the oscilloscope detects the active probe type and characteristics, and, depending on the manufacturer, it may permit the user to refine measurement and display.

Active probes differ from passive probes in other respects. Their dynamic range is lower, typically 3–8 volts. They can be damaged by probing above rated voltage and also by electrostatic discharge. To limit electromagnetic interference, the ground return lead may be screened. In addition to one megohm impedance as seen by the circuit under test, the capacitive reactance is often less than 1 pF.

The user may introduce an offset voltage to augment the limited dynamic range by adjusting the centerpoint. In this way an active probe with ±2.5 volt dynamic range can read 0–5 volts DC.

Fig. 2.26 A Keysight
1 GHz N2795A active
probe. (Keysight)

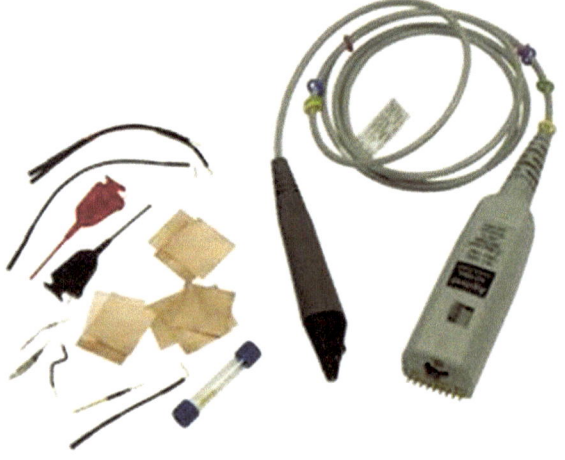

Differential Probes

Differential probes, shown in Fig. 2.27, are also active devices because they contain semiconductors, but the overall design, configuration with respect to the oscilloscope, and ultimate purpose are unique.

Let us say you want to measure and display in a standard bench-type oscilloscope the voltage between two terminals both of which are referenced to but float above the electrical system ground potential. This cannot be done using a single-ended probe because there will be a powerful fault current flowing through the ground return lead into the oscilloscope and through the power cord's green grounding conductor and then the branch circuit's bare grounding conductor to the neutral bar in the entrance panel at the electrical service. This fault current flows not when the probe tip contacts a floating voltage (that is OK) but when the ground return lead contacts a floating voltage, regardless of whether the probe tip is touching anything and regardless of whether the oscilloscope is powered up.

This problematic situation occurs when the user attempts to measure floating voltages in a three-phase Y-configuration or in either of the two ungrounded three-phase delta electrical conductors. Similarly, it arises in switching power supplies and the DC bus in a variable-frequency motor drive.

There are several strategies for avoiding the hazard. One of them is to use differential probes as opposed to a standard passive probe. Differential probes, because they are optimized to acquire only the differential signal between two test points, always reject (i.e., do not measure or display) any common mode signal. To ensure and maximize this common mode rejection, the two signal paths must be identical in terms of attenuation, frequency response, and time delay.

Differential signaling uses two complementary opposite polarity signals to convey data, each on its own conductor, from transmitter to receiver. By twisting the

Fig. 2.27 Differential probes and accessories. (Tektronix)

two conductors, as shown in Fig. 2.28, the harmful effect of a capacitively coupled external signal is diminished because it is coupled equally to the two opposing conductors, cancelling the interference.

Early differential probe designs consisted of a pair of passive probes terminated at a differential amplifier located as a separate device adjacent to the oscilloscope. An alternate design placed the differential amplifier inside the probe body, but in the age of vacuum tube amplifiers, this was unwieldy. Currently, solid-state miniaturization permits the entire circuitry to be located into a reasonably compact differential probe body, which is cabled to a single analog channel input at the oscilloscope. Since there is only one signal path outside of the probe body, costly signal path matching is reduced.

There is provision for a ground return lead to be connected to the differential probe body, but this circuit is not required for most measurements (Hickman 2001).

Included with most differential probes is a variety of hardware, which facilitates probing closely spaced test points such as adjacent pins having differing spacings in integrated circuits. The whole thing is quite user-friendly once you get past the differential probe compensation process, which is outlined in the user manual.

To begin, connect the differential probe to one of the analog channel inputs and turn that channel on. Use the Y-lead adapter to connect to the probe compensation terminals, a square-wave source on the oscilloscope front panel that is used also for compensating passive probes. Then press Autoset to display a stable square wave.

To continue the differential probe calibration, the oscilloscope should be warmed up for about 20 minutes, depending on the ambient temperature. In the Tektronix MDO3000 oscilloscope, from the Utilities menu, select Calibration. Pass should be displayed in the Status Box. If Pass is not displayed, back up and do a Signal Path Compensation. For this, disconnect all probes and signal sources from the channel

Fig. 2.28 Twisted conductor pairs, as in Cat 5e UTP cable, are widely used in Ethernet media to preserve signal integrity. (Judith Howcroft)

inputs and select Calibrate. When Pass appears in the Status Box, once again connect the differential probe to the square wave source. From the vertical menu, select Probe Calibration. In the Probe Setup window, select Clear Probe Calibration and Calibrate Probe. The differential probe calibration is complete.

Following calibration, the differential probe is ready for use. The probe is protected against static voltage. But care must be taken to prevent the probe tip amplifier from damage from over voltage.

The common-mode signal range refers to the maximum voltage with respect to ground that can be applied to either input without saturating the probe input circuitry. To test this common-mode voltage, set the probe range to 42 volts, connect a probe input to ground, and connect the other probe input to one of the differential signal outputs. Perform the test separately with each differential signal output.

Because of the electrical environment within which they are used, differential probes in some applications are required to have higher-voltage ratings, typically 600 volt. The price range is around $350 to over $5000. These probes come in various sizes, the smaller ones characterized by higher bandwidth. The larger sizes, while less portable, have higher-voltage ranges. With 30-foot cable, the larger probe can be placed at some distance from the oscilloscope. The probe housing is sometimes filled with dielectric oil for high-voltage stability.

Current Probe

The oscilloscope is essentially an auto-ranging voltmeter, but with the use of a current probe, shown in Fig. 2.29, it can be configured to read current flowing through a conductor.

Like the electrician's Amprobe, the oscilloscope current probe has a pair of jaws between which a conductor can be inserted. The current probe measures the

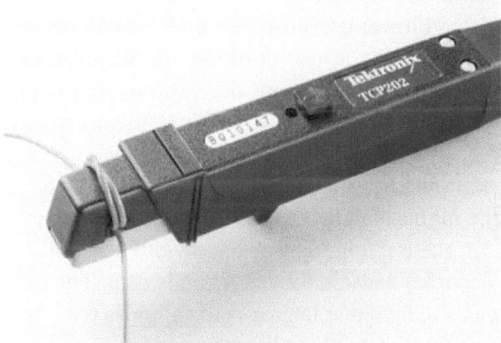

Fig. 2.29 Tektronix current probe. A technique for multiplying a weak signal is shown. Two or more turns are coiled, so they all pass through the jaws. This multiplies the current by the number of turns. The amount of current as measured by the oscilloscope must then be divided by the number of turns to arrive at the correct figure. (Tektronix)

fluctuating voltage across a magnetic coil, and the oscilloscope displays the waveform calibrated to read current. Hall effect technology is used to measure direct current.

An interesting capability of the current probe is that very weak current flow can be measured by looping the conductor two or three times around outside of a jaw and through the opening so that the surrounding magnetic field and resulting current reading are doubled or tripled.

Bandwidth

The concept of bandwidth pervades discussions of signal processing, telecommunications, test and measurement instrumentation, broadcasting, and especially oscilloscope technology (Horowitz and Hill 2015).

Bandwidth has no single definition. It varies with the context. A good starting place is frequency modulation. In its most fundamental form, FM consists of a carrier, typically sine wave, which is modulated by a much lower, often audio, signal. If you look at a time-domain representation of the modulated waveform, you will see that the result is a periodic signal whose frequency varies over time between the sum and the difference of the higher-frequency carrier and the lower-frequency signal that modulates it. The spectral separation between these two frequencies defines the bandwidth of the modulated carrier.

Another definition of bandwidth is the speed of data propagation through a network in bits per second.

The banner specification in oscilloscope vendors' literature is also called bandwidth, and this refers to the highest frequency signal that can be displayed without unacceptable attenuation. Accordingly, the amount of tolerable attenuation must be included in the overall specification for it to be meaningful. The lowest frequency at which the input signal is not attenuated in excess of 3 dB is considered by many vendors to be the instrument's bandwidth.

Generally, 1 GHz and lower oscilloscopes have a Gaussian frequency response. They exhibit roll-off beginning at one third the −3 dB level. Oscilloscopes with a bandwidth rating higher than 1 GHz typically exhibit what is known as maximally flat frequency response, flatter in the in-band portion with relatively abrupt roll-off at −3 dB.

In comparison with less than 1 GHz bandwidth instruments with Gaussian frequency response, the higher-bandwidth oscilloscopes are capable of more accurate measurements within their rating. Lower-bandwidth oscilloscopes in contrast are capable of greater accuracy outside the rated bandwidth. The rationale for attenuating out-of-band signals to a greater degree is to suppress the higher-frequency components that would cause undesirable aliasing.

The amount of bandwidth required in an oscilloscope is calculated differently for digital and analog applications. In digital work, the oscilloscope's minimum bandwidth should be five times the fastest clock speed that will be encountered. This rating is necessary to measure and display the fifth harmonic without excessive attenuation. In measuring high-speed rising edges, still greater bandwidth is needed. That is because rising (and falling) edges constitute higher-frequency components. These rise times are available in device data sheets.

Oscilloscope bandwidth required to measure and display analog signals was traditionally taken to be in excess of three times the signal frequency. But to make really accurate analog measurements, the oscilloscope must be used in a flat portion of the frequency band. Regardless of stated bandwidth, the amount of attenuation will vary for different oscilloscopes in the lower portions of the frequency band under investigation.

An oscilloscope's bandwidth is constrained by the entire signal path from probes and cable at the channel inputs, through front-end signal conditioning, analog-to-digital conversion, and subsequent processing, memory, and display. (The BiCMOS SiGe bandwidths can extend beyond 30 GHz.)

Oscilloscope Front End

Throughout the remainder of this chapter, the oscilloscope block diagram shown in Fig. 2.30 will serve as a road map. Keep in mind, however, that different types of makes and models differ significantly, so no single block diagram is universally applicable.

(The following information applies to the Tektronix TDS6804B oscilloscope with 8 GHz bandwidth and to the Tektronix TDS6154C oscilloscope with 15 GHz bandwidth.)

The oscilloscope front end may be defined as the portion of the instrument's internal circuitry from the channel inputs to the ADC. To be technically precise, we may want to place the boundary inside the ADC, at the point where samples of the analog signal are extracted.

There are separate and distinct signal paths, circuitry, and ADC's for each of the channels. When two or more channels are operated simultaneously, it is often the case that the signals at the oscilloscope inputs differ in frequency and amplitude, so they must be prepared and digitized separately prior to proceeding to the processor.

The oscilloscope front end, for each channel, is required to perform these functions:

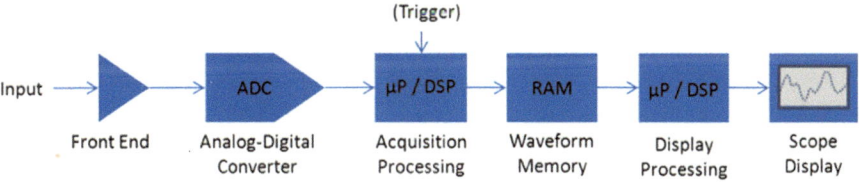

Fig. 2.30 A typical digital storage oscilloscope block diagram. (Tektronix)

- Scale the signals at the channel inputs to the ADC input ranges.
- Perform offset.
- Provide impedance conversion to prepare the signal for the ADC.
- Perform filtering where needed.
- Introduce triggering for periodic signals.

The ADC has a limited range, so the front end must condition the signals, even after they are attenuated by the 10:1 passive probe, so that they do not extend beyond that range. The first step is a scaling process. After the channel input is the attenuator, one for each channel. The ADC has no intrinsic ability to attenuate or amplify the signal. All this work has to be accomplished by the series of components upstream from the ADC. An ADC typically has a 500-millivolt range. Any greater amplitude will not be sampled, resulting in loss of information or clipping at either the high end or the low end of the ADC output.

Beyond the attenuator in the front end, for each channel there is a preamplifier. It is needed because often the attenuator, which is a stepped as opposed to continuous device, cuts the signal down to an extent where it does not fill out or make full use of the ADC half-volt range. The preamplifier, which has a user-controlled adjustable gain, tweaks the signal backup to the ADC full capacity. For very small signals that are not beyond the ADC range, the attenuator will allow the signal to pass through unmodified so that in conjunction with the preamplifier, a redundant operation is not performed.

There are many different attenuator designs. The choice is based on bandwidth requirements and signal amplitude. For up to the mid-megahertz range, a surface-mount mechanical relay is sufficient. Moving up to several gigahertz, a solid-state GaAs MESFET is used. A coaxial stepped attenuator is good for still higher frequencies, where inductive and capacitive parasitics abound. Engineers battle these challenges in direct proportion to bandwidth requirements, which translate to ultimate cost to the user. The instrument front end is an area where these factors are critical.

A similar situation exists with regard to the preamplifier. SiGe BiCMOS is standard where the bandwidth is less than 8 GHz. To 30 GHz, the package has been redesigned to limit parasitics, while beyond 50 GHz, heterojunction bipolar technology has emerged.

The signal under investigation may consist of a high DC voltage with a small oscillating signal riding on it. When this is the case, there is really no point in imposing the heavy DC burden on the ADC. It is a simple task to subtract out the DC voltage. This is done manually by turning the position knob that is near the channel button. That subtracts the DC voltage before it gets to the preamplifier or ADC. Then the user can turn the scale knob so that the AC waveform is easier to see. There are a number of ways this is accomplished, sometimes by means of an op-amp. One way or another, the DC voltage is subtracted out so that the small AC can be scaled up.

Another action that takes place in the front end is impedance selection, which is determined by the user. Pressing the channel button, the second selection in the channel menu in the Tektronix MDO3000 oscilloscope is Termination. By means of the associated soft key, the user can select 1 megohm or 50 ohm. This choice depends upon the source impedance, bandwidth, and voltage level of the signal. The probe that is used also plays a role. Most of the time, 1 megohm impedance is used, and that is the default. Occasionally 50 ohm is better as when working with high-frequency signals.

Beyond the Preamplifier

Hardware filters located upstream from the ADC perform bandwidth limiting. This is effective in limiting noise, as we have seen, because noise is a broad-spectrum and therefore high-frequency phenomenon. It does not work, of course, when the signal of interest falls in the portion of the spectrum that is filtered out. Software filtration in the processor also performs this function.

The analog signal is then filtered to provide trigger functions – low-frequency reject, high-frequency reject, and AC coupling. To achieve accurate placement of the waveform relative to the time axis, comparator outputs go to the trigger system, where they measure the time of the rising or falling edge relative to the ADC sample clock. On the oscilloscope front panel in the triggering section, the trigger level knob manually sets the threshold of the trigger comparator.

Processing Section

From the probe and analog channel input to the sampling location in the ADC, the oscilloscope is an analog instrument. Then, abruptly, it becomes entirely digital. We enter a strange new world where signals and waveforms no longer exist as originally constituted but instead are represented as digital words. As such, they can be more easily measured, transformed, analyzed, held in memory or stored indefinitely, and displayed outside of real time.

In an earlier chapter, by way of background, we discussed the ADC inner workings, particularly the different types of sampling and Nyquist requirements for valid digitation. So now we'll look beyond the ADC and examine processing and memory in a modern digital storage oscilloscope.

Digital signal processing is the decisive factor in all kinds of electronics. Following the digital revolution of the 1980s, digitation expanded beyond computer technology and emerged wherever complex electronic equipment incorporated microprocessors. TV was slow to join the trend because receivers and transmitters had to get on the same page, but at present digital signal in TV is universal.

The oscilloscope front end as described above remains an analog domain because many signals, human-made and in nature, are analog and they have to be preconditioned before reaching the ADC with its very limited dynamic range. As for digital signals, they are already digitized, and as such they can bypass the front end and ADC. That is why the relatively high digital channel count does not influence the cost of an oscilloscope to the extent that the analog channel count does.

Unlike analog technology, digital signal processing is more independent of component tolerance variation, temperature drift, and aging. *It is what it is* in the same sense that digital audio disc rendition is not easily distorted. Moreover, complex processing algorithms are easily implemented. Examples are averaging, interpolation, math functions, measurements, and display processing. (This is a very partial list.)

Leaving the ADC, the digital signal feeds into the demux, which is the gateway to acquisition memory. The demux, an IC, performs multiple functions. Foremost among them is to decelerate the flow of samples from the ADC and place them in acquisition memory. It may send out these samples in batches of 16 or 32. Other demux functions are bandwidth enhancement filtering on a software level to augment hardware filtering we saw in the front end, peak mode acquisition, and trigger positioning.

The acquisition memory is distinct from the main system processor memory. (In the oscilloscope digital section, there are three separate memories: acquisition memory, main system processor memory, and display memory.) It is only after the acquisition is complete that the main system processors can perform operations on the digital data. Additionally, data may be returned to the demux IC for additional processing.

Bandwidth Enhancement

(The following information is applicable to the Tektronix TDS6804B oscilloscope with 8 GHz bandwidth and the Tektronix TDS6154C oscilloscope with 15 GHz bandwidth.)

Data rates are accelerating well into the gigabit/s range. Optical communication is now over100 Gb/s, with 1 Tb/s a reality. RF wireless communications broadband signals are mid-GHz. To meet channel data capacity and regulatory mandates, RF and optical communications use complex modulation schemes and low-amplitude signals. Very high-bandwidth oscilloscopes are required to debug these designs. As a result, instrument manufacturers have had to extend oscilloscope performance to above 70 GHz.

The oscilloscope user has the option to enable bandwidth enhancement where desired. The purpose of this software filter is to extend bandwidth and thereby flatten channel frequency response, improve phase linearity, improve match among channels, decrease rise time, and improve time domain step response. Bandwidth enhancement improves eye diagrams, which are problematic in some less robust oscilloscopes. Bandwidth enhancement is useful for phase and magnitude measurements in the frequency domain as well as in the time domain.

It is appropriate to turn off the bandwidth enhancement filter where there is an impedance mismatch between the device being tested and the oscilloscope channel input. The user also has the option of providing custom software to filter and interpolate data.

Bandwidth enhancement is effective in removing ringing that is due to Gibbs phenomenon. This occurs at what is known as a jump discontinuity such as the upper corner of a periodic square wave rising edge. The partial sum of the Fourier series fluctuates greatly at this vicinity, and bandwidth limiting is useful in controlling those harmful oscillations.

Bandwidth enhancement promotes accuracy in measuring and comparing signals in more than one channel because during manufacture, the individual channel filters are matched in phase and magnitude response due to the fact that they are higher-frequency phenomena than the original waveform.

Digital Oscilloscope Memory

Digital oscilloscopes have a limited amount of acquisition memory. Analog oscilloscopes had no memory at all (Hickman 2001). The memory depth sets the amount of waveform time and the number of serial packets that the instrument can capture at any given sample rate. The user can slow down the time/div setting to capture longer time spans and a greater number of serial packets. But as soon as the oscilloscope's maximum time span at its maximum sample rate has been exceeded, the instrument will automatically reduce its sample rate. Then it cannot provide accurate waveform detail due to its inherent bandwidth and maximum sample rate. As we have seen, segmented memory can improve this situation.

By dividing the oscilloscope's available acquisition memory into smaller segments and overlaying successive captures while the instrument is in an infinite persistence mode, a larger periodic waveform than otherwise possible can be captured and displayed (reducing the possibility of missing a rare anomaly).

Techniques to Increase Bandwidth

As manufacturers have increased the frequency range of their electronic equipment, oscilloscope makers have of necessity worked to increase the bandwidth of their instruments in order to remain relevant. In this they have been highly successful as bandwidths have soared to ever-new multi-GHz heights. The aspiration has been to achieve these higher spectral levels without introducing unwanted distortion. Methods for doing so have been controversial, each of the three big manufacturers emphasizing different solutions and sometimes disparaging their competitors' solutions. The primary methods are:

- Interleaving (previously successfully used in the ADC and now extended into the processing section to increase overall bandwidth)
- DSP boosting, consisting of separating out by means of frequency filtering those higher-frequency components that would otherwise be disproportionally attenuated and amplifying them to compensate
- Raw hardware, consisting of continuously redesigning signal path components, circuits, and subsystems so as to reduce parasitic capacitive and inductive reactances

Interleaving

LeCroy has favored the technique known as interleaving. They call it random interleaved sampling (RIS). It is an acquisition mode that produces waveforms at very high effective sample rates. In a LeCroy oscilloscope, the user can select RIS mode from the time base settings menu when the time base is set at or below 20 μs.

RIS employs a hardware device known as a time-to-digital converter (TDC). It places the trigger position between samples. This device charges at a uniform slope between the start of the trigger to the next sample. Following that, it discharges with a uniform, lower slope. This provides an exceptionally fine temporal resolution.

The value read from the TDC is used to bin successive acquisitions. Then, multiple acquisitions from the waveform and TDC values are used to prepare acquisitions for interleaving to form a single high sample-rate waveform.

Since the waveform results from multiple acquisitions, it follows that the interleaved waveform segments have been acquired at widely separated points in time. It does not matter when the original waveforms were acquired. The order in which they are interleaved is random.

Because of the random nature of the waveforms, the waveform must be repetitive, and in each sweep, the trigger event must be the same. There can be no variation in the periodic waveform. The analog waveform must be triggered at precisely the same location at each acquisition. Any non-repetitive event, such as jitter, will invalidate the RIS process.

RIS is useful in signal integrity measurements where a TDR pulse or similar stimulus is applied to the circuit under investigation. It is also useful in device characterization where precise timing and threshold crossings are required. RIS does not work for jitter, noise measurements, and eye patterns.

DSP Boosting

Tektronix has favored the technique known as DSP boosting. To gain added bandwidth, the high-frequency portion of the signal that rolls off near and beyond the bandwidth limit is amplified to compensate for the amount it is attenuated as a result of high reactance losses.

To improve oscilloscope channel response, a DSP arbitrary equalization filter is used. It extends the bandwidth, flattens the oscilloscope channel frequency response, improves phase linearity, provides better match between channels, decreases rise time, and improves time domain step response. The user controls when bandwidth enhancement is enabled. The filter is useful in:

- Rise time measurement
- Comparison of signals in multiple channels
- Overshoot and undershoot measurements
- Spectral analysis
- Eye diagrams

Raw Hardware

Keysight has favored a technique they call raw hardware performance. The basic idea is to achieve true analog bandwidth by means of hardware innovations that eliminate parasitics and improve physical performance with respect to bandwidth. For example, new semiconductors such as preamplifier and trigger chips are introduced, and also the manufacturer must create designs that will put these new devices to work. Keysight has been acting on the theory that interleaving and DSP boosting involve trade-offs that undermine the quest for ever-greater bandwidth. Keysight states that as a result of its true raw hardware analog bandwidth, their 9000 X-series oscilloscope has the exact same noise density from 1 to 2 GHz as it does from 31 to 32 GHz (the industry's lowest noise density). It is also able to achieve an extremely flat frequency response over the full 32 GHz range. This same raw hardware chip innovation is used in Keysight active probes so that they don't constitute a bottleneck.

Keysight states that frequency interleaving requires more (not better) hardware along the signal path, which sets the stage for added noise and less accurate measurements.

In regard to DSP boosting, Keysight contends that it has a significant drawback, which is that when the signal is amplified so is the noise of the oscilloscope. Again, increased noise means less accurate measurements.

Interpolation

Following sampling, which is performed within the ADC in conjunction with the clock, usually external to it, the waveform consists of a sequence of points that can be located in two-dimensional space, referenced to an X- and Y-axis. These points can be displayed, but a more useful waveform is created when the points are joined by a line to display a continuous curve. This process is known as interpolation.

Interpolation should be first understood as a mathematical concept. It is the process by which unknown values are calculated and inserted between known values so as to form a continuous sequence. The four most used types of interpolation are interpolation off (dots only), pulse, line, and $\sin(X)/x$.

In dots only, that is without interpolation as shown in Fig. 2.31, only the original sample points of the waveform are displayed. If the sample rate is high and time/div is long, the sample points will almost merge to form a nearly continuous trace, and interpolation may not be needed.

Figure 2.32 shows the same Gaussian waveform with $\sin(X)/x$ interpolation.

$\sin(X)/x$ uses a short sine wave segments to connect each pair of display points. The amplitude of this curve depends upon a number of factors, primarily the distance between the points.

Pulse interpolation draws a horizontal line from the first display point to the time of the next display point. Then it finishes the interpolation by drawing a vertical line to the next display point. This creates a step pattern, which is the approximate shape of an ideal waveform.

Interpolation can be properly implemented only when sampling is fully compliant with the Nyquist theorem, which states that a signal can be perfectly reconstructed from discrete samples only if these rules are observed:

- The highest frequency component sampled must be less than half the sampling frequency.
- Samples must be acquired at equal intervals.

It is important to observe, in the first of these requirements, that the signal that must be less than half the sampling frequency is the *highest* frequency component.

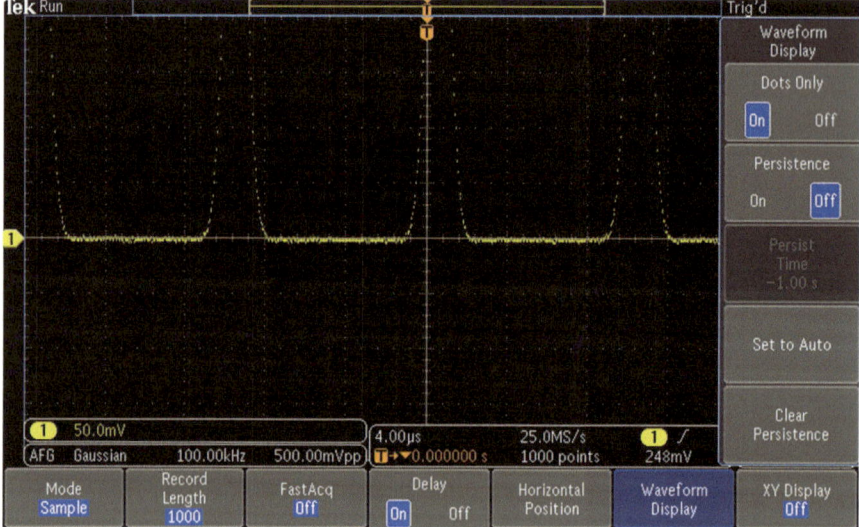

Fig. 2.31 Dots only display of a Gaussian waveform. (Author's screenshot)

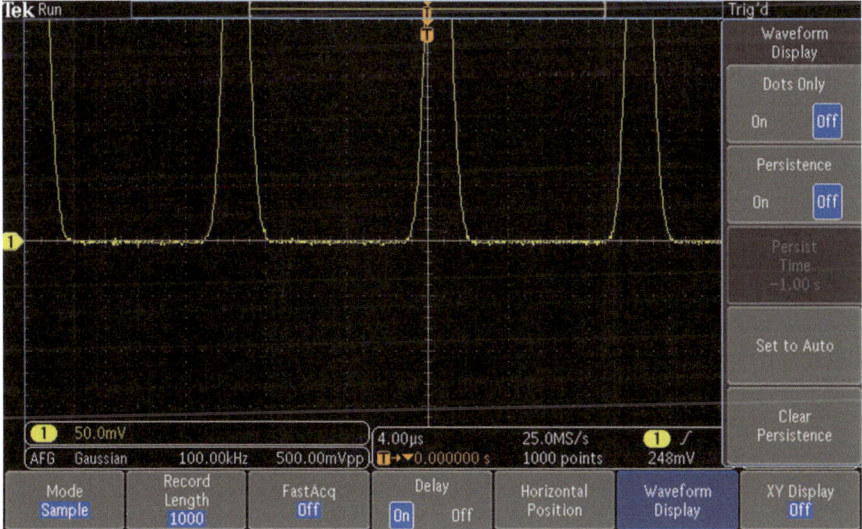

Fig. 2.32 Sin(X)/x interpolation produces a more coherent display. (Author's screenshot)

Since a non-sinusoidal wave is composed of the fundamental plus higher-frequency harmonics, a faster sampling rate is required compared to a simple sine wave. That is why, for example, the very fast rise and fall times of a square wave are imperfectly sampled when the fundamental only is used to calculate the required sampling rate.

Signals can be reconstructed by various methods, but precision can be attained by using sin(X)/x, provided the two Nyquist conditions are observed.

Oscilloscope Memory

In an oscilloscope there are separate memories. With respect to the processor, acquisition memory is upstream, and display memory is downstream. In speaking of memory, it is usually the acquisition memory that is referred to. Because this memory holds samples from the ADC output, its size is critical. Its size is called "memory depth." That term should not be taken to imply that the acquisition memory has some sort of vertical architecture. Deep memory simply means large memory. Memory depth, sample rate, and time captured are related as follows:

$$\text{Sample rate} = \text{Memory depth} / \text{Time captured}$$

The sample rate is a banner specification for any digital oscilloscope, and as such it is usually prominently displayed on the front panel along with the bandwidth. For a 1 GHz bandwidth instrument, for example, a typical sample rate is 5 GS/s (five billion samples per second!) This is the maximum sample rate. The actual value varies according to the above equation and is subject to change according to the user-controlled time captured.

The number of samples that can be stored in the acquisition memory depends upon the memory depth. A deep memory permits the user to maintain the maximum sample rate regardless of time base setting. This facilitates accurate measurements. In debugging, it is important to see infrequent and brief transients, and for this high resolution is critical. An oscilloscope that has insufficient memory depth will lack the resolution required to zoom in and see signal details.

The user can easily set record length. In a Tektronix MDO3000 oscilloscope, press Acquire with or without a signal displayed. Then, in the horizontal Acquisition menu at the bottom of the screen, press the soft key associated with record length. Turning Multipurpose Knob a, the user can choose among six record lengths, ranging between 1000 points (samples) and ten million points. At a specific sampling frequency, the record length is the number of samples that is taken in the duration of the measurement. The total measuring time is determined by the record length, as set by the user. If you increase the record length, more of the signal is displayed in accordance with this equation:

$$\text{Measurement duration in seconds} = \frac{\text{Record length in samples (points)}}{\text{Sampling frequency in Hz.}}$$

The maximum available memory depth depends upon an oscilloscope's maximum sample rate. Oscilloscopes sample at close to the fastest rate when the time base is adjusted to a faster range. Setting the time base to a slower range in order to display longer time spans causes the oscilloscope to automatically reduce the sample rate based on available memory. The relevant equation is:

$$\text{Acquisition Memory} = \text{Time Span} \times \text{Sample Rate}$$

Contrary to first impression, deep memory is not always better. The problem is that acquiring long waveforms using deep memory equates to increasing waveform processing time, which means that waveform update rates are reduced so that the oscilloscope is less responsive, and as a result in debugging, transients may be lost. That is the rationale for adjusting memory depth by changing the record length.

The amount of memory required to store an entire signal as captured by the oscilloscope depends upon the time duration of the waveform and the sample rate. The time duration of the waveform is determined by the user. It is the horizontal scale setting, which is adjusted by turning the scale knob in the horizontal section. Note that there is a single pair of horizontal knobs, position and scale, for all channels, while there is a separate pair of knobs for each channel in the vertical section. What that means is that for each channel, the amplitude can be separately adjusted, but time remains synchronized for all channels.

For a specified sample rate, the memory requirement becomes larger as the time duration increases. To deal with this problem, the sample rate is reduced, not in the

ADC but in the processor. This is accomplished by choosing the correct sampling mode, as discussed previously in this text. In the default sampling mode, one sample is saved in memory during each sample interval, regardless of the ADC sampling rate as determined by the clock. Excess samples are discarded, i.e., not saved in memory. This is known as "decimation." The downside is that for debugging where we are looking for an anomaly, it may be thrown away.

The user, by pressing mode in the Acquisition menu, can change the sample mode from the default, where samples are discarded, to Peak Acquisition Mode. Here the highest and lowest peaks from adjacent pairs of sample intervals are saved in the acquisition memory, and the problem of lost samples and lost signal anomalies is avoided.

Tektronix MDO3104 Front Panel Controls

Many of the buttons as shown in Fig. 2.33 have diverse functions or serve only to open a horizontal menu that appears across the bottom of the display. The Multipurpose Knobs are contextual, performing multiple tasks depending on the menu item that is active. When only a single menu item is to be selected, Multipurpose Knob a is designated for the task. If two menu items are to be activated, both Multipurpose Knobs are used. For this reason, Multipurpose Knob a gets much more use than Multipurpose Knob b.

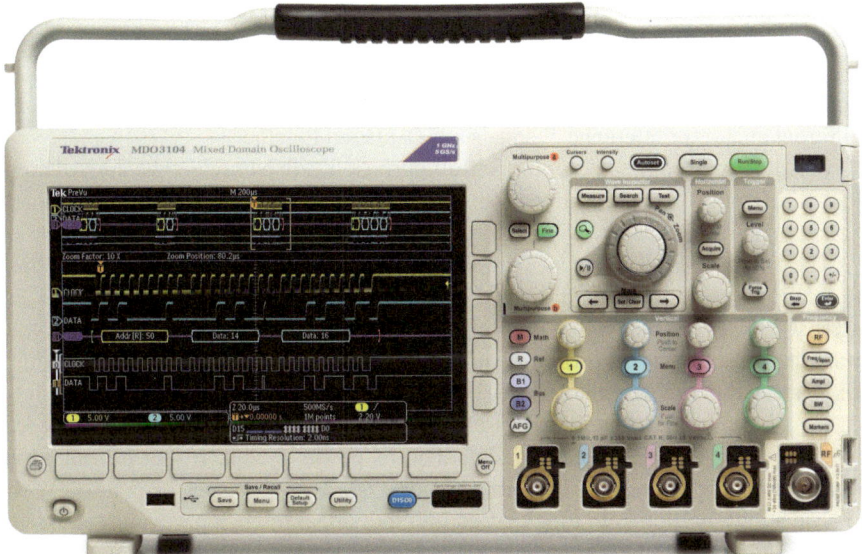

Fig. 2.33 Tektronix MDO3104 front panel knobs and buttons. (Tektronix)

Vertical and horizontal rows of soft keys activate menu selections. They are "soft" in the sense that they vary depending upon the operation that has been chosen by the user. The major oscilloscope manufacturers conform to this model and terminology. Here's a rundown:

When a front panel button is pressed, the oscilloscope displays a menu for the desired function. If the Math button is pressed while one or more of the analog input channels is active, the Math menu appears across the bottom of the screen. Available menu selections are Dual Waveform Math, FFT, Advanced Math, Label, and Auto-Scale. Below each selection is an associated soft key.

Pressing the soft key associated with Dual Waveform, a vertical submenu appears to the right of the display. Each item has an associated soft key. Dual Waveform Math performs any of the four basic arithmetic operations, add, subtract, multiply, and divide, on any of two waveforms that are present at the analog inputs or reference channels. These are labeled first source and second source and may be chosen by turning Multipurpose Knobs a and b. The operator, add, subtract, multiply, or divide, may be chosen by pressing the associated soft key. Then the math waveform appears, in its dedicated red color.

If there is no input at an analog channel port (no BNC cable or probe), there is still a signal present, with zero amplitude. Accordingly, the math operations can be and are performed. Since you can't divide by zero, that display has a strange appearance.

FFT is the next selection in the horizontal math menu. Pressing the associated soft key brings up the vertical FFT menu. Only a single input is used, one of the four analog channel inputs, or one of the four reference channels. It is selected by turning Multipurpose Knob a. Then, pressing Menu Off, the user can obtain a clearer view of the display. Time domain and frequency domain displays are both shown. In an MDO instrument, they are shown in split-screen format, which is why the oscilloscope is called mixed domain. The same signal is shown in two domains as opposed to in an MSO or mixed-signal oscilloscope, where separate signals from different analog or reference channels are displayed together.

In Advanced Math, the user turns Multipurpose Knob a to select the desired analog or reference channel, followed by pressing the soft key associated with Enter Selection and OK Accept.

The widely used Reference button opens the horizontal Reference Menu. Reference waveforms may be created or recalled and displayed to set up a standard against which to compare other waveforms or for other purposes such as when you want to preserve a signal for future reference. Waveforms may be entered with time and date stamped. When they are recalled, they are displayed in the dedicated reference color, white. These reference waveforms are nonvolatile, which means they are retained when the oscilloscope is defaulted or power-cycled. If the instrument is recalibrated, the reference waveforms are lost.

The horizontal soft keys toggle on and off individual reference waveforms. When one or more is on, a vertical side menu appears. Then vertical and horizontal settings may be adjusted, labels applied, reference details acquired, and waveforms may be saved to external storage such as a flash drive inserted into a USB slot.

The front panel button labeled AFG when pressed displays in any analog channel or in the RF Channel One of the arbitrary function generator waveforms stored in the oscilloscope. A BNC cable is run from AFG Out on the rear panel to the desired channel input on the front panel. With this channel on, an AFG waveform is chosen after pressing the soft key associated with waveform in the lower horizontal AG menu. Then, pressing waveform settings, frequency/period, amplitude, and offset may be selected.

The analog channel buttons permit the user to turn on one or more channels. When pressed, a horizontal menu appears across the bottom of the screen. Menu items are coupling, termination, invert, bandwidth, label, fine scale, offset, position, probe setup, and deskew.

Coupling may be AC or the default DC, which is used most of the time. (A more appropriate label would be AC/DC.) It displays the AC waveform plus any DC component. AC eliminates the DC component. It is used when the DC is relatively high so that the AC component cannot be accurately displayed. AC is used to accurately display and measure ripple in the output of a DC power supply.

Termination is one megohm or 50 ohm. One megohm is the default, and it is used almost all the time. It causes the oscilloscope to resemble a high-impedance multimeter, thereby not loading the circuit under investigation. The 50-ohm termination is used with a low-amplitude signal when necessary to match impedance in order to get an accurate reading.

Invert is normally left off. When turned on, it reverses the signal's polarity. The effect may be seen by toggling it with a waveform displayed.

Bandwidth is normally kept at full, the default. Lowering it temporarily reduces the bandwidth so as to limit noise, a high-bandwidth phenomenon. When finished, the oscilloscope should be defaulted.

The Acquire button is very important. It brings up the horizontal Acquisition Menu, comprised of several selections. Pressing the soft key associated with mode permits the user to choose the sampling mode. Sample, the default, is used most of the time. When necessary, peak detect and high resolution use sample points that would be discarded in sample mode. They operate only when the sample rate is below the maximum. The sample rate is determined by setting the horizontal scale and record length.

Record length sets the number of points or samples. It is adjusted by turning Multipurpose Knob a and may be one thousand, ten thousand, one hundred thousand, one million, five million, or ten million.

FastAcq selects the way in which the waveform palette is displayed. Multipurpose Knob a determines whether it is shown as temperature or spectrum, normal, or inverted.

Delay may be toggled on or off. Its effect may be observed by accessing a signal and observing the effect as delay is toggled. The effect of horizontal position may be observed by pressing the associated soft key and turning Multipurpose Knob a. Position and scale are quite different for all channels in the horizontal section as opposed to individual channels in the vertical section.

Waveform display, dots only, and persistence can be turned on or off. Persist time may be set by turning Multipurpose Knob a, or it may be set to automatic.

XY display may be turned off (the default) or set to triggered XY, displaying the Lissajous pattern.

The Trigger Menu button permits the user to select triggering options. Types include edge, sequence (B trigger), pulse width, time-out, runt, logic, setup and hold, and rise-fall time. Trigger coupling may be DC, AC, high-frequency reject, low frequency, and noise reject (DC low-frequency sensitivity). Mode may be auto or holdoff. Holdoff inhibits triggering until a user-chosen amount of time elapses, and another triggering event occurs. Its purpose is to prevent false triggering if there is an anomalous spike.

Wave Inspector

Wave Inspector is controlled by the buttons and knobs in the upper center part of the front panel, which has two large concentric knobs and eight buttons. One purpose of Wave Inspector is to permit the user to get a different perspective on long record length waveforms. Of the two concentric knobs, the outer one pans the signal, and the inner one zooms the signal. To the left of them, labeled with a magnifying glass icon, is the zoom button. Pressing it repeatedly toggles zoom off and on. When it is on, the button is lighted, and an overall view of the waveform appears as a separate display in split-screen format at the top.

In this upper display, a pair of white brackets indicates the area shown in the lower view. The zoom knob moves these brackets farther apart, for a minimum 2× zoom factor, or closer together for a maximum 100 zoom factor. The area between the brackets is shown in the bottom display. That way, the user can obtain a very detailed view of the waveform under investigation. Similarly, the outer pan knob moves the pair of brackets from side to side, permitting the user to search for anomalies.

The play-pause button automatically pans the waveform. By turning the outer pan button, the speed can be regulated. Pressing the set button in the mark section, a small white triangle marks for future reference the current location of the brackets. Pressing the button a second time clears the marker. The horizontal arrow buttons move the brackets to existing marks.

Search marks can automatically be set and cleared by pressing the search button above the concentric knobs. User-defined marks are solid, while automatically set marks are open. The soft key associated with search in the horizontal menu at the bottom toggles on and off the vertical search menu at the right. The top soft key toggles search on and off. Soft keys can be pressed to clear all marks, copy search settings to trigger, copy trigger settings to search, convert automatic marks to user marks, and turn the time- and dated-stamped mark table on and off. The Wave Inspector includes these search capabilities:

- Edge – Searches for edges (rising, falling, or both) with a user-specified threshold level.
- Pulse width – Searches for positive or negative pulse widths that are >, <, =, or ≠ a user-specified pulse width or are inside or outside of a range.
- Timeout – Searches for the lack of a pulse. The signal stays above or below (or either above or below) a set value for a set amount of time.
- Runt – Searches for positive or negative pulses that cross one amplitude threshold but fail to cross a second threshold before crossing the first again. Search for all runt pulses or only those with a duration >, <, =, or ≠ a user-specified time.
- Logic – Search for a logic pattern (AND, OR, NAND, or NOR) across multiple waveforms with each input set to either high, low, or don't care. Search for when the event goes true, goes false, or stays valid for >, <, =, or ≠ a user-specified time. Additionally, you can define one of the inputs as a clock for synchronous (state) searches.
- Setup and hold – Search for violations of user-specified setup and hold times.

Measure

Also in the Wave Inspector section is a button labeled Measure. Pressing it opens a horizontal menu at the bottom. Then, pressing the soft key associated with Add Measurement opens the vertical Add Measurement submenu at the right. Multipurpose Knob a selects the source, and Multipurpose Knob b selects the measurement type, such as frequency, period, rise time, fall time, etc. Additionally, the user can select Snapshot All. When the selections have been made, pressing OK Add Measurement causes the measurement to be placed in the display, which can be read if Menu Off is pressed.

The fourth soft key activates the on-screen digital voltmeter. In the vertical submenu, Multipurpose Knob a selects the type of measurement, which is displayed at the top of the screen. Choices are off, AC + DC RMS, DC, AC RMS, and frequency, as shown in Fig. 2.34.

As previously discussed, Waveform Histograms, Statistics, Gating, and Reference Levels can be activated.

Test

Defaulting the oscilloscope and pressing the Test button in the Wave Inspector section, the horizontal Test menu appears at the bottom. Pressing the soft key associated with Application, a vertical submenu comes up on the left, permitting the user to select Limit/Mask Test, Video Picture, Power Analysis, or Act on Event. If Multipurpose Knob a is used to select Power Analysis and the soft key associated with Analysis is pressed, the vertical Power Analysis submenu appears on the right.

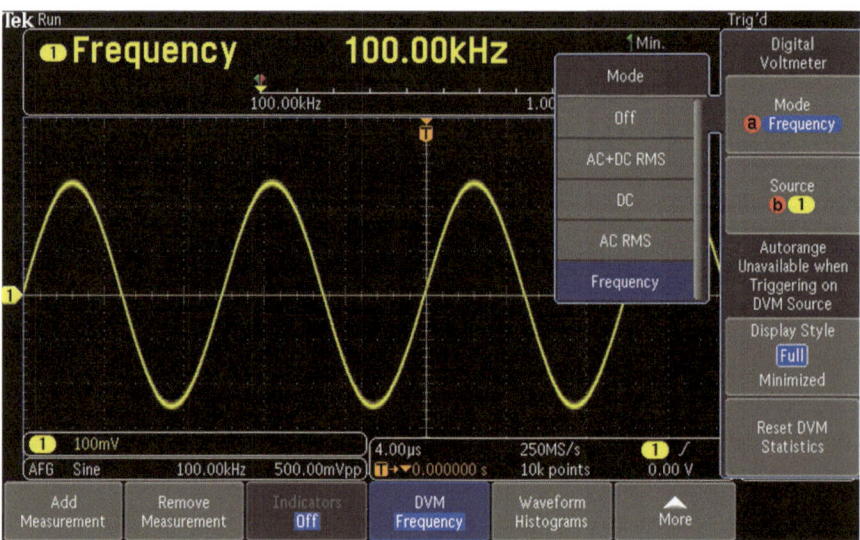

Fig. 2.34 In DVM mode, frequency is displayed. (Author's screenshot)

Choices are power quality, switching loss, harmonics, ripple, modulation, safe operating area, and deskew. Each of these categories causes a horizontal submenu to appear, opening a window that provides further details.

References

Gibilisco, Stan, *Reading Schematics*, Third Edition, McGraw-Hill, 2014
Hickman, Ian, *Oscilloscopes*, Fifth Edition, Newnes, 2001
Horowitz, Paul and Hill, Winfield, *The Art of Electronics*, Third Edition, Cambridge University Press, 2015
Gibilisco, Stan and Monk, Simon, *Electricity and Electronics*, Sixth Edition, McGraw-Hill, 2016

Chapter 3
Oscilloscope as a Diagnostic Tool

Abstract Bench-top vs. hand-held, battery-operated instruments, and use of the differential probe. The danger of short circuit when a bench-type instrument is used to measure a voltage referenced to and floating above ground level. Hazards in servicing a variable frequency drive (VFD) used to control a motor, particularly in measuring and viewing the DC bus voltages. While the differential probe is effective, most engineers and technicians use the hand-held, battery-operated oscilloscope with inputs insulated from each other and from ground. Working with oscilloscope front-panel controls. Fast Acq, Wave Inspector, and acquisition modes. Sampling in the digit oscilloscope. Types of interpolation.

Noise is for the most part unwanted acoustic, electrical, or physical energy that interferes with a useful signal and places a lower limit on the amplitude of any signal that can be received or measured (Horowitz and Hill 2015). A stronger signal may be received in a form that is compromised by the presence of noise at or near the same frequency. Noise can be an intrinsic part of the signal, interference from an outside source or measuring instrument, or it can be generated within the measuring instrument including probe or cable connecting to that instrument to the device under investigation.

The definition of noise always involves a subjective assessment. That is because, aside from human judgment, noise is not different from a signal of interest. Noise has no practical use except in random number generation, as an element in the creation of dither and for educational demonstrations.

There are mitigation techniques that can reduce or eliminate noise, and we'll describe them in this chapter. They are not always effective at high levels. Noise can damage electronic instrumentation and pose a health threat.

Types of Noise

Principle types of noise include but are not limited to thermal noise, shot noise, flicker noise, burst noise, and interference. These varieties differ in frequency and amplitude distribution, and consequently if rendered in acoustic form, they sound

© Springer Nature Switzerland AG 2020
D. Herres, *Oscilloscopes: A Manual for Students, Engineers, and Scientists*,
https://doi.org/10.1007/978-3-030-53885-9_3

different, and if visually displayed in an oscilloscope, they may be distinguished from one another.

Thermal noise, also called Johnson or Nyquist noise, is a prominent phenomenon. Any resistive object including naturally occurring as well as manufactured devices exhibits thermal noise. It may be seen in the output of a semiconductor, which is a varying resistance. (Noise at the input of an amplifier is subject to gain, which increases the noise amplitude.)

Thermal noise can be measured by means of a voltmeter. You may have noticed that a digital multimeter, when in volts mode, displays a low-level fluctuating voltage. This is sometimes called phantom voltage, mysterious to apprentice electricians. It is a manifestation of thermal noise.

Voltage across any resistance that is not in a circuit other than created by the high-impedance measuring instrument consists of the random motion of charge carriers. Both amount of motion and measured voltage are directly proportional to the temperature of the resistive body. This thermal noise, at a reasonable frequency, has a flat spectrum and, because it is analogous to light, is considered a form of white noise.

The equation that relates thermal noise, also called Johnson or Nyquist noise, to absolute temperature is:

$$V_{\text{noise}} = \left(4kTRB\right)^{1/2}$$

Where k = Boltzmann's constant

T = temperature in degrees Kelvin
R = resistance in ohms
B = bandwidth in Hertz

What is Boltzmann's constant and why is it in this equation? It is a *very* small number, by convention denoted k, equal to $1.38064852 \times 10^{-23}$. This constant relates the average kinetic energy of particles in a gas to the temperature of that gas.

It is also equal to the gas constant R divided by that Avogadro constant N_A:

$$k = R / N_{A}$$

This important constant, discovered by Ludwig Boltzmann (1844–1906), is relevant in any discussion of ways in which properties of atoms (mass, charge, structure) relate to properties of matter (viscosity, thermal conductivity, diffusion).

Thermal noise plays a role when instrumentation, especially the oscilloscope, is used to measure or display faint electrical signals. It is interesting to note that noise plays a role in electrical signals, but not in the corresponding mathematical functions, which are too abstract to exhibit noise.

If the leads connected to a resistor's terminals are shunted, a measurable current will flow through the circuit that is formed. The amount is:

$$I_{noise} = \left(4kTB / R\right)^{1/2}$$

Where B – Boltzmann's constant

The amplitude of thermal noise voltage and current is unpredictable at any given instant, but it conforms to a Gaussian distribution curve, and accordingly it can be measured. We don't have to worry about the very faint signal being lost below the noise floor because it *is* the noise floor, as shown if Fig. 3.1.

There are types of noise other than thermal. They are quite different in source and form:

- Shot noise is characterized by a succession of discrete, temporally separate, electrical charges. Like thermal noise, shot noise is white (broad spectrum) and Gaussian. The term "shot noise" is appropriate. Converted to and audible signal and suitably scaled, it sounds like rain on a steel roof.
- Flicker noise differs from thermal noise and shot noise (which are theoretical properties of any resistance) in that it arises from manufacturing anomalies in resistors. In a real-world circuit, all three of these noise types co-exist, and, if of sufficient amplitude, they may be measured. Flicker noise, also called pink noise, is quantified as 1/f because it inversely relates to frequency. Flicker noise depends upon the construction of a resistor, having to do with the resistive material and interfaces at terminals.
- Burst noise is comprised of discontinuous and seemingly random noise that does not conform to a Gaussian curve. It was a problem in 1970's era and earlier semi-conductors (many are still around), and it consists of random jumps between two

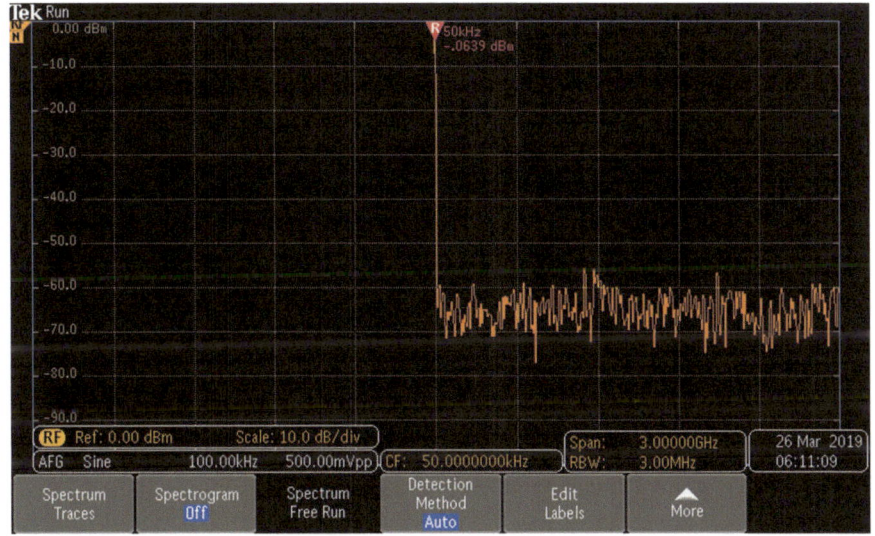

Fig. 3.1 Noise floor as it appears in an oscilloscope frequency domain display (Math>FFT) with 0 volts at the analog input. (Author's screenshot)

voltage levels. The duration is typically 1/100 of a second. In the frequency domain, burst noise appears as low-frequency events without discernible spikes.
- Band-limited noise, as the name suggests, is not a broad-spectrum phenomenon like thermal noise. Later we'll discuss mitigation techniques including waveform averaging and band-pass filtering.
- Interference, of course, comes from outside the signal and (usually) outside the measuring instrument. It is often very apparent and can be identified by the spectrum and amplitude signature. Prominent examples are 60 Hz hum and uninvited spectrally close broadcast signals. Switching power supplies, with abrupt rise and fall times, powerful motors and cell phones are frequent sources. Shielding within grounded metal enclosures is an effective remedy. If the shielding does not have a low-impedance ground, it can exacerbate the problem by acting as a parasitic element in an antenna array. Another successful mitigation technique, applicable for low-intensity, short-range interference, is to physically relocate the source, measuring instrument, circuit under investigation, and connecting cables.

Signal-to-Noise Ratio

Signal-to-noise ratio (SNR) is just what it says, signal divided by noise. A stronger signal or reduced noise equates to a higher SNR, as shown in this Equation:

$$SNR = P_{signal} / P_{noise}$$

Where P is average power. P_{signal} and P_{noise} must be measured at the same or equivalent locations, and constrained by the same bandwidth. Moreover, because these quantities are signal power over noise power, usually expressed in decibels, the relations are logarithmic. The reason for introducing decibels into the equations is that the signal (with noise) often has a large dynamic range. Thus:

$$P_{signal,dB} = 10\log_{10}\left(P_{signal}\right)$$

$$P_{noise,dB} = 10\log_{10}\left(P_{noise}\right)$$

Also:

$$SNR_{dB} = 10\log_{10}\left(SNR\right)$$

Applying the logarithmic quotient rule:

$$10\log_{10}\left(P_{signal} / P_{noise}\right) = 10\log_{10}\left(P_{signal}\right) - 10\log_{10}\left(P_{noise}\right)$$

Finally:

$$SNR_{dB} = P_{signal,dB} - P_{noise,dB}$$

All of that is much simpler than the series of equations suggests. It is just the way a decibel scale works.

Two Noise Mitigation Methods

Because noise is a broadband phenomenon, whose frequency extends far above the upper frequency of the impacted signal, it is often possible to clean up the display (increase the SNR) by a technique known as bandwidth limiting. Modern DSO's incorporate means for reducing the bandwidth. Press the applicable analog channel button. Then press the soft key associated with bandwidth. The Bandwidth menu permits the user to set the bandwidth to different levels, as shown in Figs. 3.2, 3.3, and 3.4.

It is important to realize that the signal (and noise) bandwidth is reduced, not the bandwidth of the oscilloscope. Of course bandwidth reduction will not work where the signal bandwidth is above the cut-off frequency.

Another method for mitigating noise that is obscuring a signal of interest is waveform averaging, also known as signal averaging. It enters the oscilloscope acquisition picture because the better-known technique, bandwidth limiting,

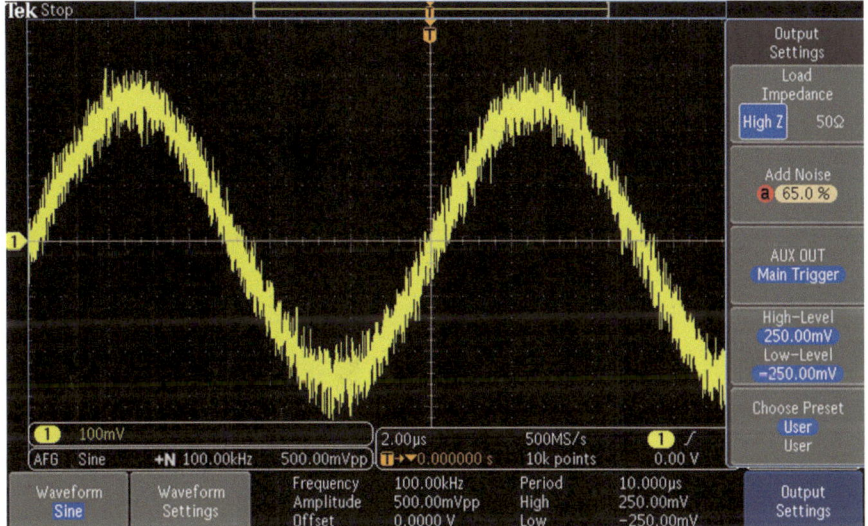

Fig. 3.2 shows a sine wave at full bandwidth with 65 percent noise added. (Author's screenshots)

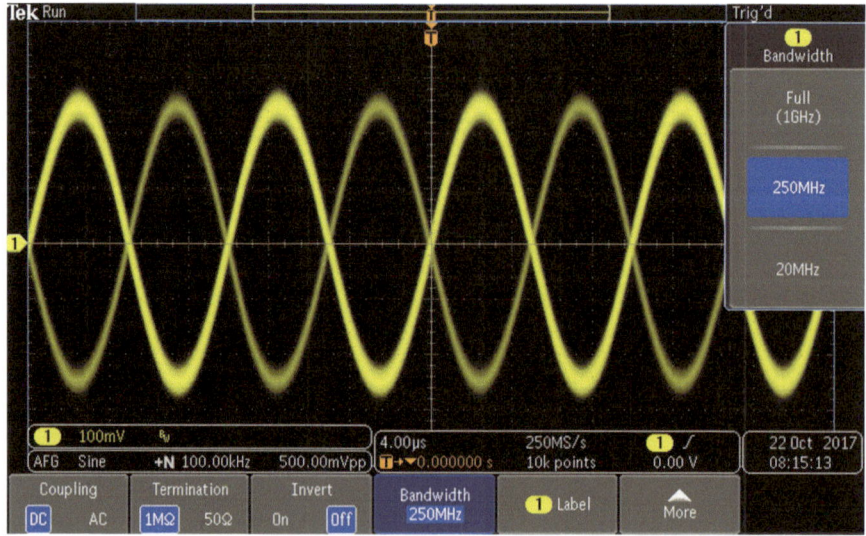

Fig. 3.3 shows the same signal with bandwidth reduced to 250 MHz. (Author's screenshots)

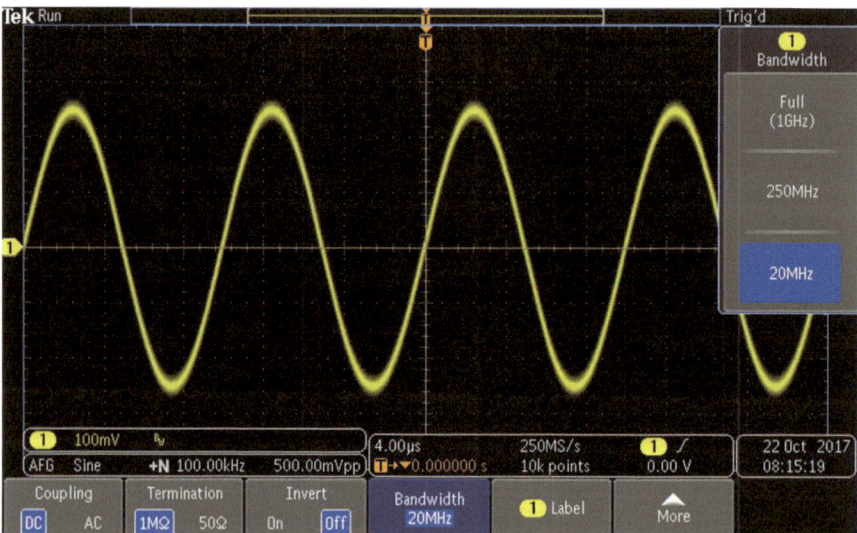

Fig. 3.4 shows the same signal with bandwidth reduced to 20 MHz. Notice the noise reduction. (Author's screenshots)

does not work well with high-frequency signals. Waveform averaging, an alternate method for strengthening the signal of interest relative to noise, works well on high-frequency signals that are repetitive or periodic. In this method, a number of signals, specified by the user, are averaged to produce a composite signal. Because thermal noise is random as opposed to cyclic, the averaging process surresses it while strengthening the signal under investigation.

To see waveform averaging in action, press AFG. With a sine wave displayed, press Output Settings and add 20 percent noise, as shown in Fig. 3.5.

Then press Acquire>Mode. In the Acquisition Mode menu, press the bottom soft key, associated with Average. Using Multipurpose Knob a, the number of wave-forms that are averaged can be set from a minimum of two to a maximum of 512. As you increase the number, the sine wave as displayed with Noise Added regains triggering and the noise goes away, as shown is Figs. 3.6 and 3.7.

More About Noise Floor

A critical concept in any discussion of oscilloscope or other measuring instrumenta-tion is the noise floor. If noise is defined as any signal other than the one(s) under investigation, the noise floor is the sum of noise and unwanted signals that are pres-ent in the measuring instrument circuitry. This may be partly internal and partly external. (What is a signal of interest from the point of view of one instrument may be noise to another measuring instrument. A human perspective is sometimes

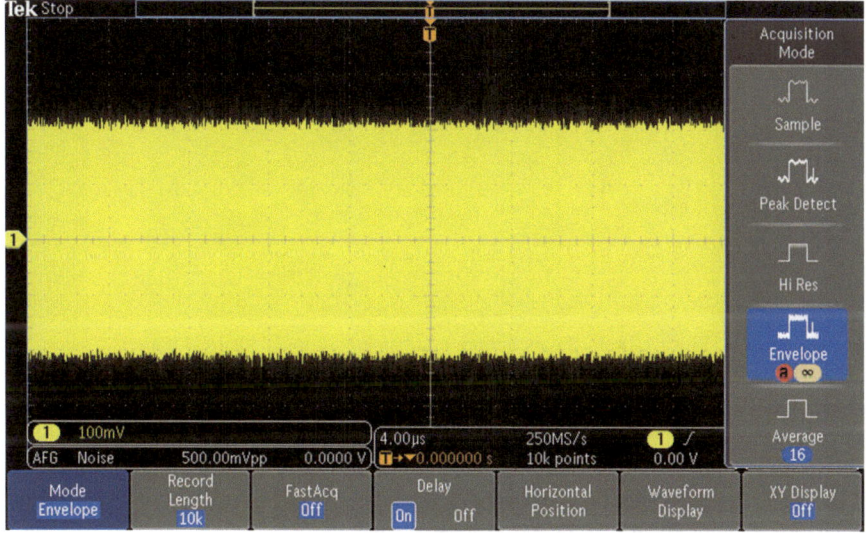

Fig. 3.5 Noise added to signal. (Author's screenshot)

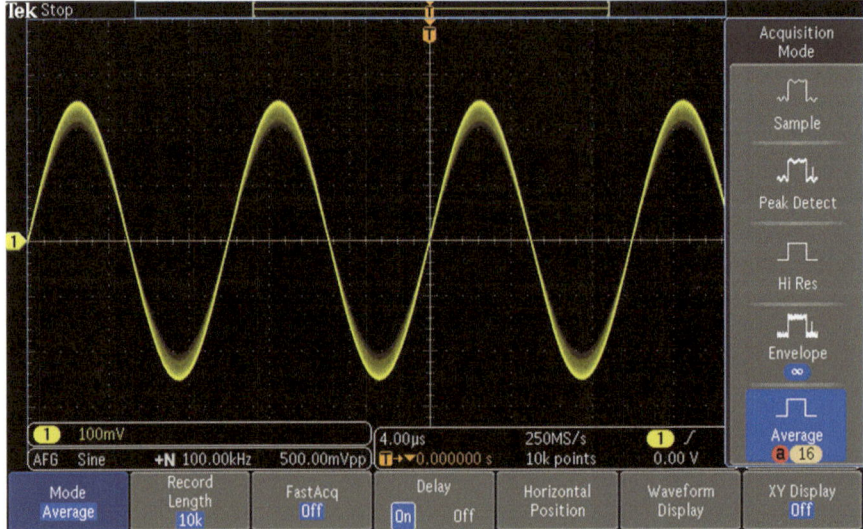

Fig. 3.6 When 16 waveforms are averaged, the noise component is diminished. (Author's screenshot)

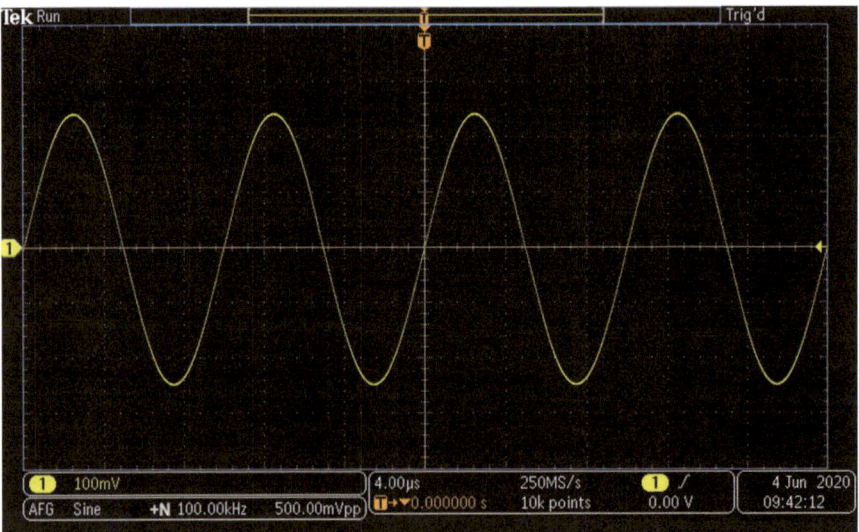

Fig. 3.7 When 512 waveforms, the maximum number, are averaged, thermal noise is substantially eliminated. (Author's screenshot)

needed to make the final determination, though usually it is obvious.) Probe cables, power lines, and any nearby conductive bodies can function as unintended antennas. The only way to gain total control over the external portion is to build a Faraday cage, which is an enclosure used to block electromagnetic fields. The internal thermal noise, within the oscilloscope of measuring instrument, can be suppressed by building ever higher-tech and more expensive instrumentation, but noise can in principle never be totally eliminated.

Noise sources include internal and external thermal noise, black body and cosmic background noise, atmospheric thunder, energy radiated by sunspots, and human-made noise.

Internal instrumentation noise and external and electromagnetic and physical noise combine to make a pseudo-random irregularly fluctuating electrical wave at the instrument output. Its most prominent effect is to obscure low-amplitude signals of interest and to frustrate our efforts to measure or display them. The noise floor prohibits measuring a signal whose average amplitude drops below it.

It is, however, sometimes possible to retrieve signals that would otherwise disappear below the noise floor when bandwidth limiting and waveform averaging are not effective. This is accomplished by using spread spectrum techniques, where information at a particular bandwidth is made to expand in the frequency domain so that it can successfully compete with the inherently broad-spectrum, higher amplitude noise.

Frequency hopping is an important technique that is implemented in telecommunications. It is a transmission protocol that involves changing the frequency of a carrier at a rapid rate based on a pseudorandom pattern shared by transmitter and receiver. The technique is known as frequency hopping spread spectrum (FHSS). It is a code division multiple access scheme.

In the old analog oscilloscope, the signal at the input was first amplified and then fed to the vertical deflection plates in the CRT in concert with a triggered time-base voltage fed to the horizontal deflection plates. Our present-day digital storage oscilloscope works somewhat differently. It makes use of a two-stage process: first the signal is acquired and digitized, then a processed version is applied to the flat-screen display. The processor mediates this action and has access to stored waveforms in memory, which can be recalled and analyzed in multiple ways.

The oscilloscope user can control the signal acquisition process by press the contextual soft keys and turning knobs on the front panel. Following digitalization in the ADC, the processor converts sample points into waveform display points. The all-important acquisition menu is the user's gateway for controlling the output as displayed, stored, and recalled.

The sample interval is a measure of separation between discrete samples. It is to be emphasized that sample intervals and waveform intervals are not necessarily identical. One consequence of this is that there are several acquisition modes, which determine the ways in which sample points transform into clusters of wave points.

Acquisition Modes

The simplest and most fundamental type of acquisition is the sample mode. Here the oscilloscope automatically saves one sample point for each waveform in order to assemble a display. A periodic signal is needed at the oscilloscope input for the sample mode to work.

Two consecutive waveforms are used in Peak Detect mode. The two corresponding waveform points derive from minimum and maximum values that are saved in memory. In Peak Detect mode, the ADC operates at a fast sample rate with a slow time base setting. The result is a series of long waveform intervals, enabling the processor to acquire rapid changing signals. The value to the user in this mode is that the instrument can display fast transients at slow sweep speeds when narrow pulses occur far apart.

High Resolution is similar to Peak Detect. In both of these modes, the ADC samples the analog waveform at a higher speed than the time base setting actually requires. A sine wave that is acquired in High Resolution mode retains a large amount of noise, which manifests in a thickened, blurry appearance of the trace. In Sample mode, the trace is thinner and has a sharp appearance. One advantage of High Resolution mode is that it works on any signal without regard to whether or not it is repetitive.

Envelope mode is a variation of Peak Detect mode. The difference is that in Envelope mode, the waveform that the processor installs in the oscilloscope memory is comprised of waveform points that are obtained from different acquisitions rather than from the same acquisition. This results in a radically different appearance in the display, which has to be seen, as in Fig. 3.8, to be appreciated.

Peak Detect mode uses data acquired from the oscilloscope input and from it the processor constructs Envelope mode. Setting the number of peaks at infinity, it is observed that the trace of a pure sine wave thickens greatly. In this instance, the altered appearance does not signify noise content, but instead it is a statistical variation among the constituent waveforms. As a striking demonstration, adjust the AFG to synthesize Noise and display it in Envelope mode. If you insert a range of waveform numbers in succession, the nature of Envelope mode becomes apparent.

Next in the Acquisition mode menu is Average, which can be used to good effect in clearing up a signal that has been obscured by noise. Noise elimination, by means of bandwidth limiting or signal averaging (also correctly called waveform averaging) exploits the fact that noise as an electrical phenomenon differs from the signal under investigation. Noise is random, enabling signal averaging, and it is broad spectrum, enabling bandwidth limiting. In signal averaging, a number of waveforms, specified by the user, are averaged. In this way, due to its random nature, noise is reduced if not substantially eliminated. Bandwidth limiting works by taking out the higher frequency portion of the noise spectrum, beyond the bandwidth of the signal of interest. Waveform averaging sometimes succeeds where bandwidth limiting fails because it cannot display signals outside its reduced frequency span. This can be seen when viewing a signal in the frequency domain (Math>FFT).

In signal averaging, the user specifies the number of waveforms to be averaged, ranging from 2 to 512. As you turn Multipurpose Knob a or use the keyboard to increase this number, you can watch a noisy signal regain clarity.

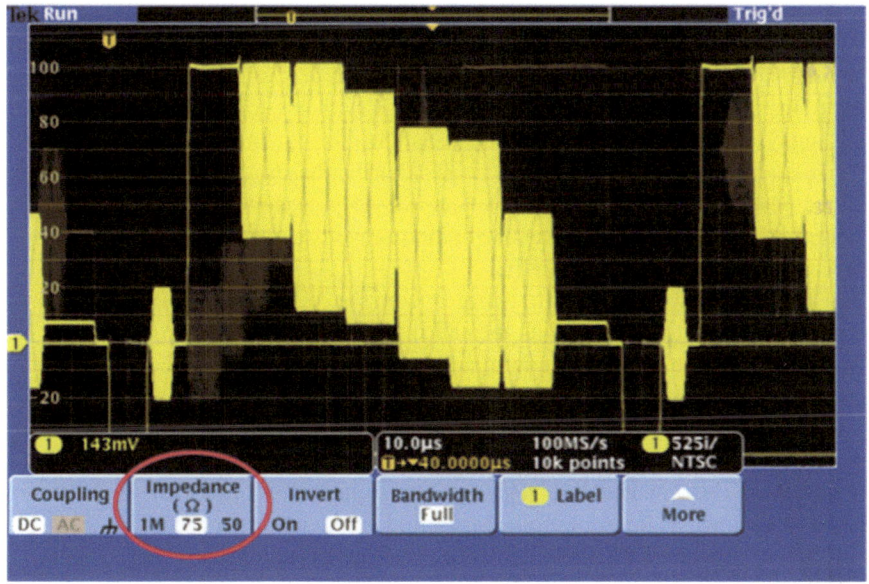

Fig. 3.8 Envelope Mode. (Author's screenshot)

The acquisition menu permits the user to turn on and off and adjust parameters other than acquisition modes. The first of these, beginning at the left, is Record Length. Multipurpose Knob a or the keypad permits the user to adjust the number of points from 1000 to 10 million.

You would think that a long record length is preferable, and this is often but not always true. Sometimes, it is necessary to reduce record length in order to optimize performance. The sampling rate is equal to record length divided by temporal length of the acquisition. Depending upon the frequency of the signal under investigation, memory size may dictate a reduction in memory length to ensure that parts of the signal are not lost.

A very interesting oscilloscope function, the next menu item, is FastAcq, which can be toggled on to bring up the Waveform Palette. Options are Temperature, Spectral, Normal, and Inverted Displays. FastAcq extends the oscilloscope's ability to perform high-speed waveform capture, which is effective in revealing elusive signal anomalies. The way FastAcq does this is by reducing dead time that would otherwise occur between separate acquisitions. This permits us to see glitches and runts. The intensity of the waveforms as displayed indicates their rate of occurrence, exactly the information we need to see to evaluate how performance is impacted and to find the source(s) of harmful distortion.

The digital storage oscilloscope can store waveform information in its memory, and the acquisition can be stopped and resumed as needed, functions that would not be possible in an analog oscilloscope.

Waveform Inspector is a very user-friendly facility that allows us to see waveform details from alternative perspectives. Using the large concentric knobs and buttons on the front panel, we can examine a signal that is no longer present at the input. Instead, it is stored in memory where it is available for display, measurement and analysis. It can be zoomed, panned, searched for dysfunctional characteristics, marked, and saved. An oscilloscope acquisition can be stopped and restarted using buttons at the top of the front panel. Single Sweep, activated by a nearby button, causes the oscilloscope to display one or more waveforms as determined by the time base and then to stop automatically.

The digitized waveform as it emerges from the ADC if displayed in that form would consist of a series of dots located in the space scaled according to the X- and Y-axes. These Cartesian coordinates place the dots developed by the sampling process, but for a satisfactory display, they must be connected to make coherent traces. This process can be likened to stepping back into the analog world, where information is made available not just for the dots, but everywhere in between. The finished line segments are supplied by one of the algorithmic processes known as interpolation. Depending on the nature of the original signal, sine wave, square wave, digital pulse, etc., either of the two interpolation methods may be appropriate, linear, which is more efficient for rectilinear waveforms such as digital pulses, or sin x/x, which is preferred for curved and irregular traces.

Another acquisition tool that may be invoked at the user's option is Delay. To see it in action, press AFG (after defaulting the instrument if necessary). The sine wave appears in the display. Then press Acquire to bring up the Acquisition menu. Toggle Delay on and off and notice that the waveform shifts position with respect to the X-axis. Delay, when activated, moves the trigger point with respect to the expansion point. This maneuver allows the user to see detail that precedes the trigger point, which is helpful in finding spurious signals that interfere with proper equipment operation. The amount of delay can be modified by turning the horizontal position knob.

Another item in the acquisition menu is Waveform Display. Pressing the associated soft key, the Waveform Display menu appears at the right of the screen.

Dots only are self-explanatory, but where this gets interesting is in Persist Time. Eye diagrams can be created if the signal qualifies, and they are used for evaluating high-speed digital signals. The last item in the Acquisition menu is XY display, which was used in our discussion of Lissajous figures (Figs. 3.9, 3.10, 3.11, 3.12, 3.13, 3.14, and 3.15).

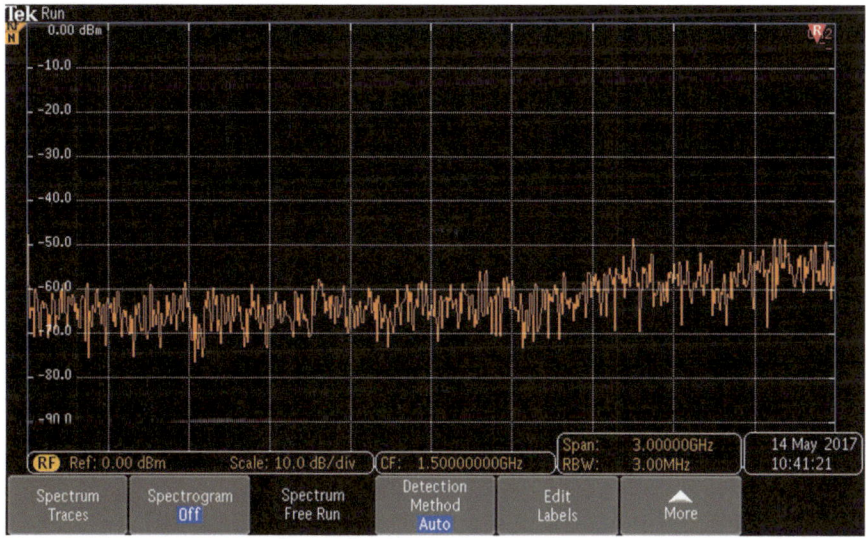

Fig. 3.9 Sine wave in frequency domain

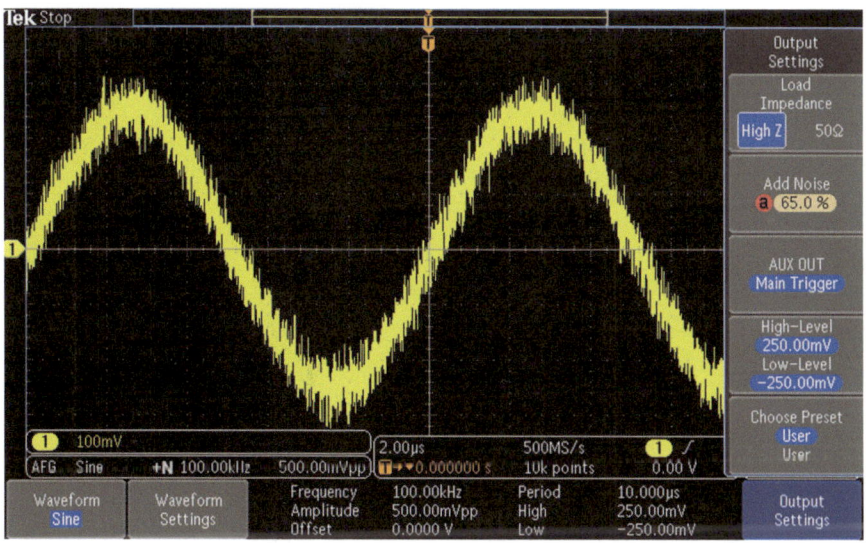

Fig. 3.10 Sine wave with 65 percent noise added

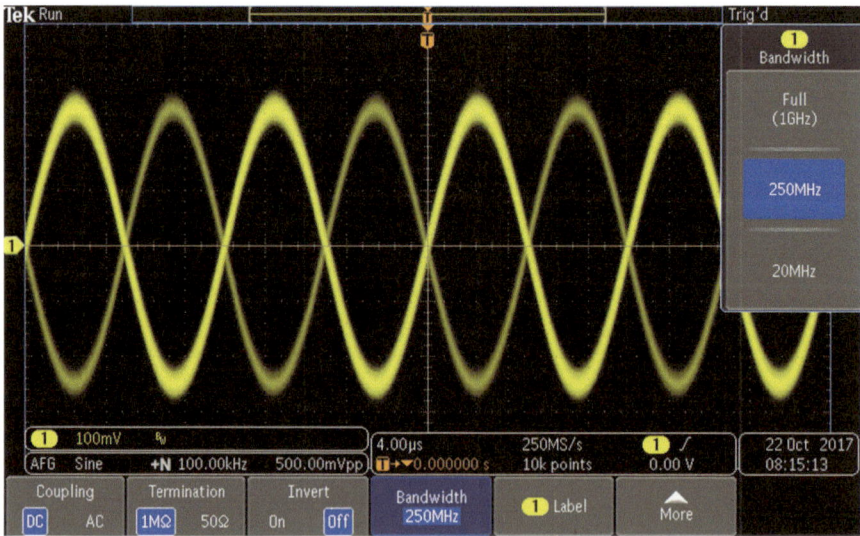

Fig. 3.11 Sine wave with 20 percent noise added

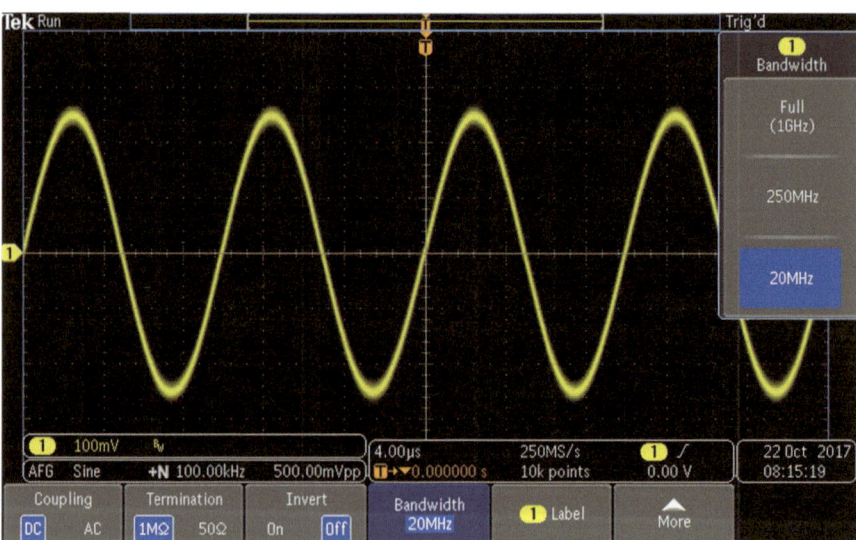

Fig. 3.12 Sine wave with bandwidth reduced to 20 MHz

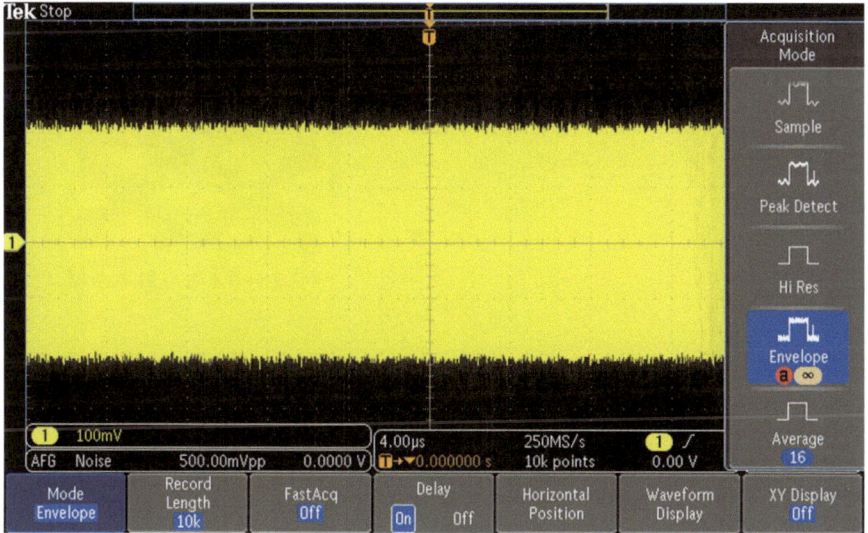

Fig. 3.13 Noise in envelope mode

Fig. 3.14 Eye diagram

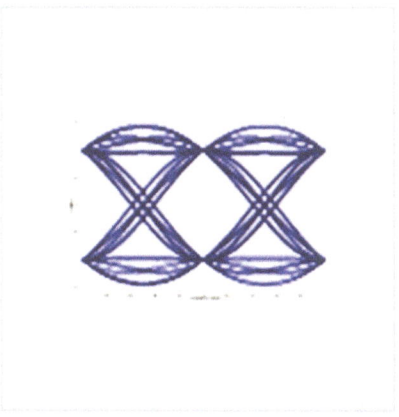

Fig. 3.15 Alternate eye
diagram

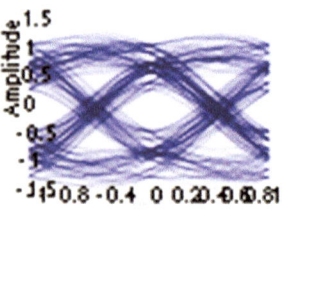

Reference

Horowitz, Paul and Hill, Winfield, *The Art of Electronics*, Third Edition, Cambridge University
    Press, 2015.

Chapter 4
Oscilloscope Math and Measurements

Abstract The math function in the latest oscilloscopes and how it can be used to gain insight into waveform relationships. In the Tektronix Series 3000 oscilloscope, pressing the Math button brings up the Math tabs. Dual Waveform Math permits the user to identify Source 1 and Source 2. Then, the Add, Subtract, Multiply, or Divide operator may be specified. Pressing the FFT button permits Source, and vertical and horizontal units to be chosen.

Mathematics is the language of science. For centuries, theoreticians have debated whether mathematics exists in an ideal domain of forms, the Platonic view, or as Aristotelians would have it, math is a mere tool developed by humans to make sense of the swirling universe. Either way, we have a compelling need to understand this complex system and learn to apply it in the vast ocean of data around us.

One facet of the oscilloscope, especially since the old-world machine morphed into the post-1980 digital instrument, is its extraordinary ability to process a signal (or a *function* in the mathematician's language). Even when the waveform is no longer live at the analog channel input but is instead a totally accurate digital image in the memory, we can recall, modify, combine it with other waveforms, measure and display the signal within Cartesian coordinates, and then return it to memory for future reference.

This chapter will be a tour of the Math and Measure functions in the oscilloscope, beginning at a basic level, the Measure mode, and proceeding through a submenu titled Advanced Math.

The Math Mode: Getting Started

Virtually all of today's digital oscilloscopes are capable of measuring and displaying a wide variety of signal parameters. We'll see how they are measured and how the metrics are displayed. First, a signal has to be connected to the oscilloscope input. This can be done in several ways. If the oscilloscope has an internal arbitrary function generator (AFG), generally available as an option, it is a simple matter to

© Springer Nature Switzerland AG 2020
D. Herres, *Oscilloscopes: A Manual for Students, Engineers, and Scientists*,
https://doi.org/10.1007/978-3-030-53885-9_4

connect it by running a BNC cable from the AFG output on the back panel to one of the analog channel inputs.

After the oscilloscope is powered up, press Default Setup. (This does not remove long-range factory and user settings such as time and date, installed firmware, etc. It does remove settings from the last work session, which could cause the instrument to behave in strange ways not relevant to the job at hand. Beginning a new task, it is a good move to press Default Setup if you don't want to lose your way down a side road.)

To continue, press AFG. This brings up the default sine wave. The AFG menu appears with highlighted menu choices, as shown in Fig. 4.1.

As always, soft keys that are below horizontal highlighted menu selections correspond to them and each opens a vertical submenu. The soft keys below non-highlighted areas across the bottom of the display are not active. In the active areas we see selected parameters – frequency, amplitude, offset, trigger location along the X-axis, time per division, number of samples per second, date, and time. These are all default settings and can be changed easily by the user. For example, pressing Waveforms opens a submenu which allows you, using Multipurpose Knob a, to change the waveform from Sine to:

- Arbitrary
- Square
- Pulse
- Ramp
- DC
- Noise

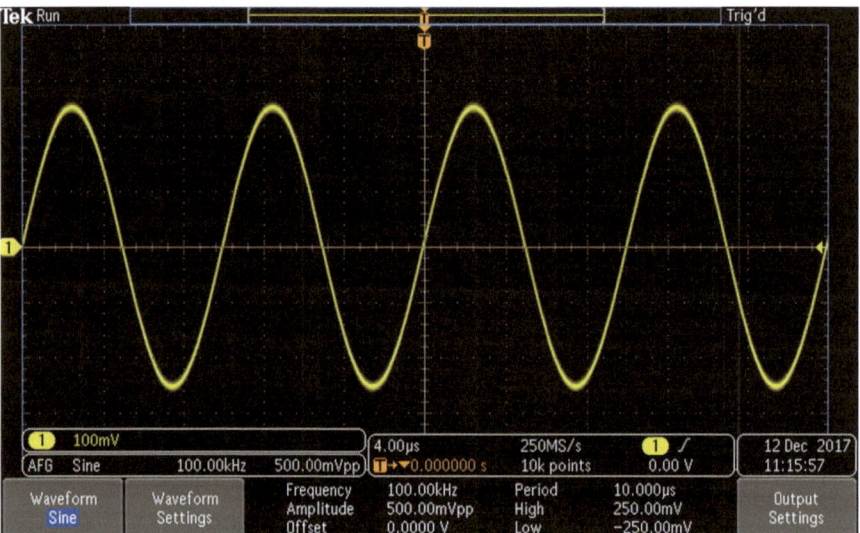

Fig. 4.1 Default AFG settings. (Author's screenshot)

- Sinx/x
- Gaussian
- Lorentz
- Exponential rise
- Exponential decay
- Haversine
- Cardiac

Similarly, pressing the soft key that corresponds to Waveform Settings permits the user, by means of Multipurpose Knobs a and b, to adjust frequency, period, amplitude, offset, high level, and low level.

There are other ways to obtain a waveform in order to demonstrate the Math mode. Of course, an internal AFG from another oscilloscope even built by a competing manufacturer can be used. Just run a BNC cable from machine to machine. Another option is an autonomous signal generator. AC utility power from a premises branch circuit is a possibility if you are aware of and can avoid potential hazards. The output from the secondary winding of a Class 2, 24-volt transformer works well. In addition to reducing the voltage, this device isolates the output circuit from the premises electrical system ground, eliminating one fault-current scenario.

When it comes to DC, which makes for some interesting Dual Waveform Math displays, a nine-volt battery is a convenient signal source. Use a 10:1 oscilloscope probe connected to an analog channel input. Also, a bench-type power supply is more than suitable.

The Measure Function

The foregoing has been in anticipation of our look at the Measure function in the digital oscilloscope. To access this mode, with the AFG sine wave displayed, in the Wave Inspector section press Measure. This action brings up the appropriate horizontal menu below the display. The menu selections are Add Measurement, Remove Measurement, DVM, Waveform Histograms, Statistics, Gating, Reference Levels, and High-Low Method. We'll talk about each of these in turn.

Pressing Add Measurement invokes a vertical submenu along the right side of the display. The top selection permits the user to set the source, using Multipurpose Knob a. The second soft key, Measurement Type, governed by Multipurpose Knob b, opens a third-level menu. Parameters that can be displayed are:

- Snapshot All
- Frequency
- Rise Time
- Fall Time
- Delay
- Phase
- Positive Pulse Width

- Negative Pulse Width
- Positive Duty Cycle
- Negative Duty Cycle
- Burst Width
- Peak-to-Peak
- Amplitude (Max, Min, High, Low)
- Positive Overshoot
- Negative Overshoot
- Mean
- Cycle Mean
- RMS
- Cycle RMS
- Positive Pulse Count
- Negative Pulse Count
- Rising Edge Count
- Falling Edge Count
- Area
- Cycle Area

Each of these waveform parameters is quantified and has an on-screen description with equations. Snapshot All, shown in Fig. 4.2, is a summary.

Each measurement type is described with an equation, diagram, and definition. For example, Frequency is shown in Fig. 4.3.

Fig. 4.2 Snapshot All displays in table form the principle metrics pertaining to the signal that is present at the chosen source, in this instance analog Channel One. (Author's screenshot)

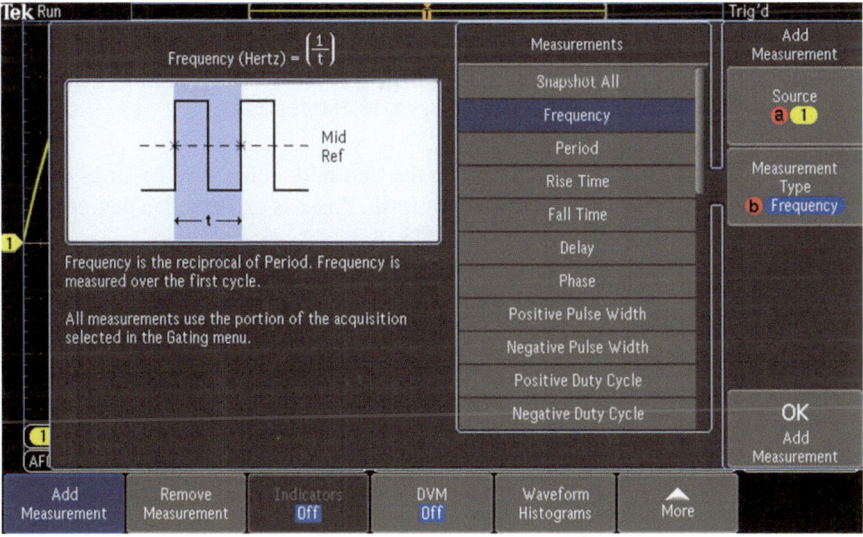

Fig. 4.3 Frequency. (Author's screenshot)

$$\text{Frequency}\left(\text{Hertz}\right) = \left(1 / \text{time}\right)$$

Frequency is the reciprocal of period. All measurements use the portion of the acquisition selected in the Gating menu.

$$\text{Period}\left(\text{seconds}\right) = t$$

Period is the time required to compete the first cycle. The first cycle is the time between the first two positive crossings, or the first two negative crossings at the mid-Reference level.

$$\text{Rise Time}\left(\text{seconds}\right) = t$$

Rise Time is the time required for the first rising edge to rise from the Low-Reference level to the High-Reference level.

$$\text{Fall Time}\left(\text{seconds}\right) = t$$

Fall Time is the time required for the first falling edge to fall from the High-Reference level to the Low-Reference level.

$$\text{Delay}\left(\text{seconds}\right) = t$$

Delay is the time between the first occurrences of the specified mid-Reference crossings of two waveforms.

$$\text{Phase}\,(\text{degrees}) = (t_1\,/\,t_2)\times 360\,\text{degrees}$$

Phase is the ratio of the delay between the first mid-Reference crossings of two waveforms to the period of the source waveform. Phase is expressed in degrees. 360 degrees comprise one cycle.

$$\text{Positive Pulse Width}\,(\text{seconds}) = t$$

Positive Pulse Width is the time between the mid-Reference crossings of the first positive pulse.

$$\text{Negative Pulse Width}\,(\text{seconds}) = t$$

$$\text{Positive Duty Cycle}\,(\text{percent}) = (t_1\,/\,t_2)\times 100\,\text{percent}$$

Positive Duty Cycle (percent) is the ratio of the Positive Pulse width to the period. Positive Duty Cycle is measured over the first cycle.

$$\text{Negative Duty Cycle}\,(\text{percent}) = (t1\,/\,t2)\times 100\,\text{percent}$$

Negative Duty Cycle is the ratio of the Negative Pulse Width to the period. Negative Duty Cycle is measured over the first cycle.

$$\text{Burst Width}\,(\text{seconds}) = t$$

Burst Width is the time from the first mid-Reference crossing to the last mid-Reference crossing.

$$\text{Peak} - \text{to} - \text{Peak}\,(\text{volts}) = a$$

Peak-to-Peak is the difference between the maximum value and the minimum value.

$$\text{Amplitude}\,(\text{volts}) = a$$

Amplitude is the difference between the high value and the low value.

$$\max\,(\text{volts})$$

Max is the maximum value.

$$\min(\text{volts})$$

Min is the minimum value.

$$\text{High}(\text{volts})$$

If the High-Low Method is Histogram, High is the highest density of points above the midpoint of the waveform. If the High-Low Method is Min-Max, High is equal to Max.

$$\text{Low}(\text{volts})$$

If the High-Low Method is Histogram, Low is the highest density of points below the midpoint of the waveform. If the High-Low method is Min-Max, Low is equal to Min.

$$\text{Positive Overshoot}(\text{percent}) = (a_1 / a_2) \times 100 \text{ percent}$$

Positive Overshoot is the difference between Max and High, divided by amplitude, as shown in Fig. 4.4.

$$\text{Negative Overshoot}(\text{percent}) = (a_1 / a_2) \times 100 \text{ percent}$$

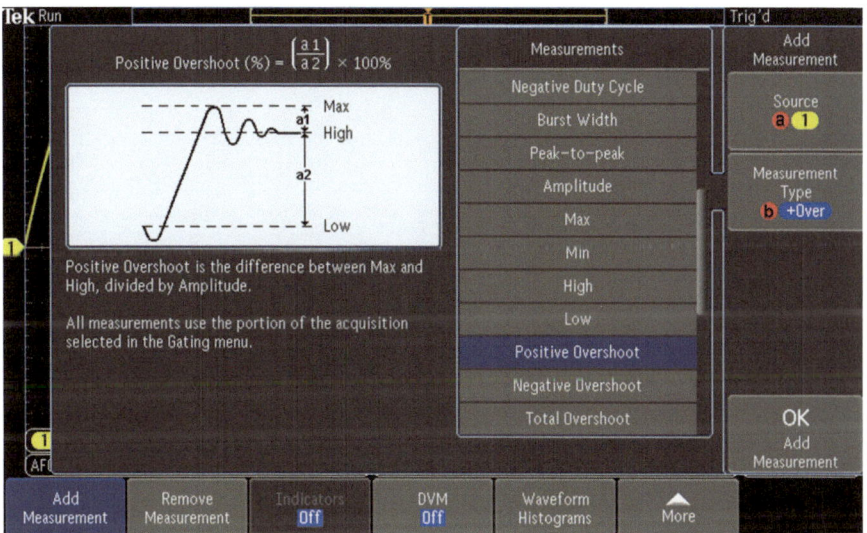

Fig. 4.4 Positive Overshoot. (Author's screenshot)

Negative Overshoot is the difference between Min and Low, divided by Amplitude, as shown in Fig. 4.5.

Total Overshoot (percent) is the summation of Positive Overshoot and Negative Overshoot.

$$\text{Mean}(\text{volts}) = (S_0 + S_1 + \ldots + S_n)/n$$

Mean is the arithmetic mean value, as shown in Fig. 4.6.

$$\text{Cycle Mean}(\text{volts}) = (S_0 + S_1 + \ldots + S_n)/n$$

Cycle Mean is the arithmetic mean value. Cycle Mean is calculated over the first cycle.

$$\text{RMS}(\text{volts}) = \text{square root of} \left(S_0^2 + S_1^2 + \ldots + S_n^2\right)/n$$

RMS is the true Root Mean Square value.

$$\text{Cycle RMS}(\text{volts}) = \text{square root of} \left(S_0^2 + S_1^2 + \ldots S_n^2 / n\right)$$

Cycle RMS, shown in Fig. 4.7, is the true root mean square value calculated in the first cycle.

$$\text{Positive Pulse Count} = n$$

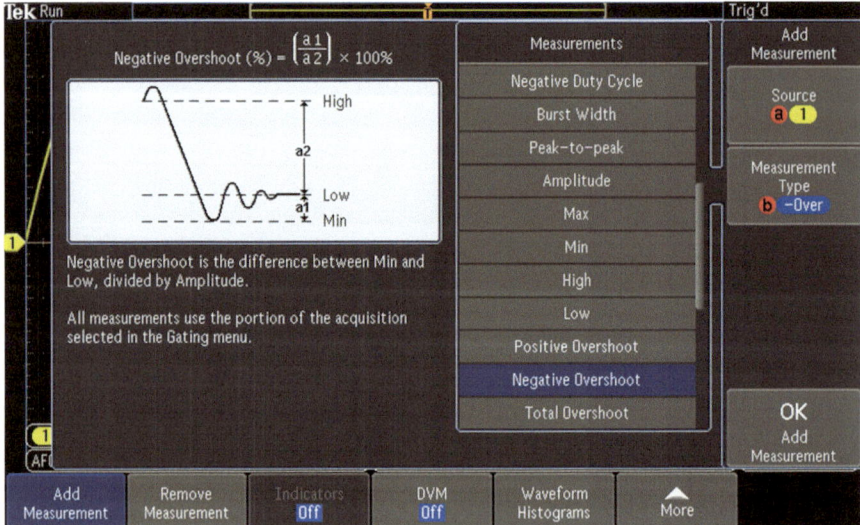

Fig. 4.5 Negative Overshoot

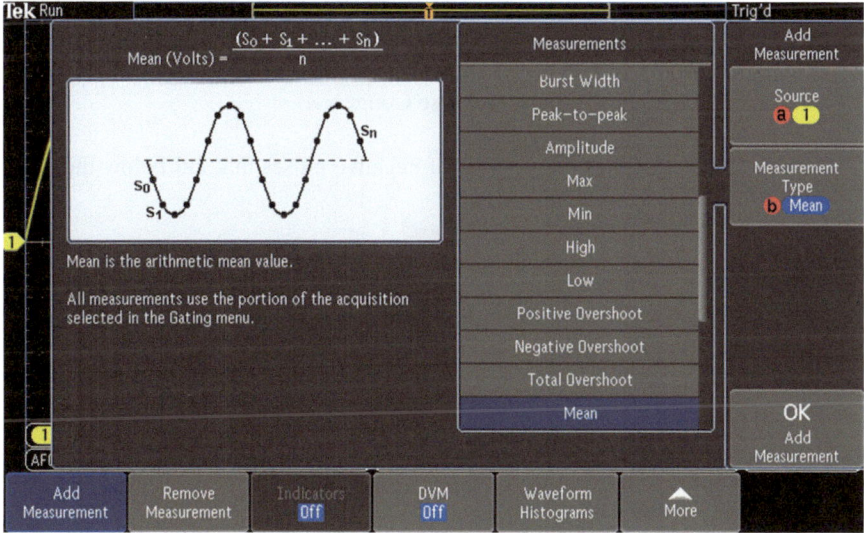

Fig. 4.6 Mean

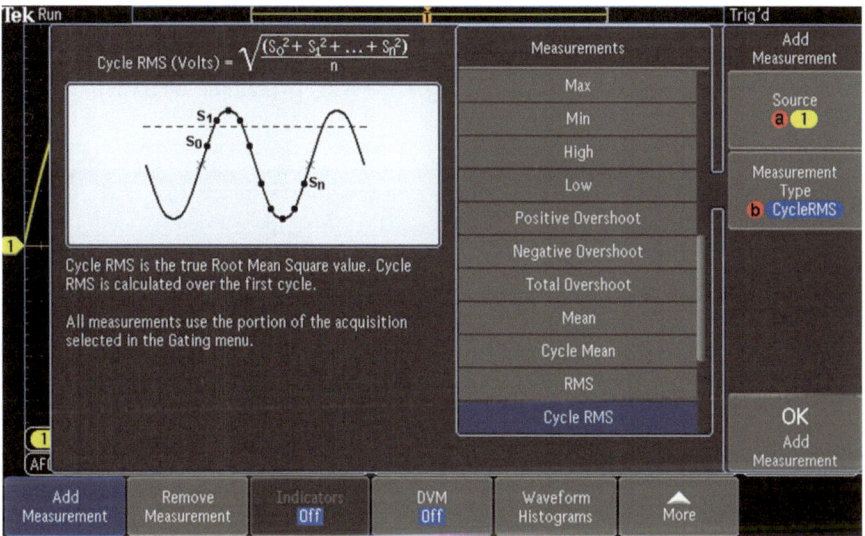

Fig. 4.7 Cycle RMS. (Author's screenshot)

Positive Pulse Count is the number of positive pulses that rise above the mid-level, as shown in Fig. 4.8.

$$\text{NegativePulse Count} = n$$

Negative Pulse Count is the number of negative pulses that fall below the mid-Reference level.

$$\text{Rising Edge Count} = n$$

Rising Edge Count is the number of positive transitions from the Low-Reference value to the High-Reference value.

$$\text{Falling Edge Count} = n$$

Falling Edge Count is the number of negative transitions from the High-Reference value to the Low-Reference value.

$$\text{Area}(\text{Volt} - \text{seconds}) = \left(S_0 + S_1 + \ldots + S_n\right) \times (\text{sample interval})$$

Area is area under the curve, calculated by integrating the samples, as shown in Fig. 4.9. The area measured above ground is positive. The area measured below ground is negative.

$$\text{Cycle Area}(\text{volt} - \text{seconds}) = \left(S_0 + S_1 + \ldots S_n\right) \times (\text{sample interval})$$

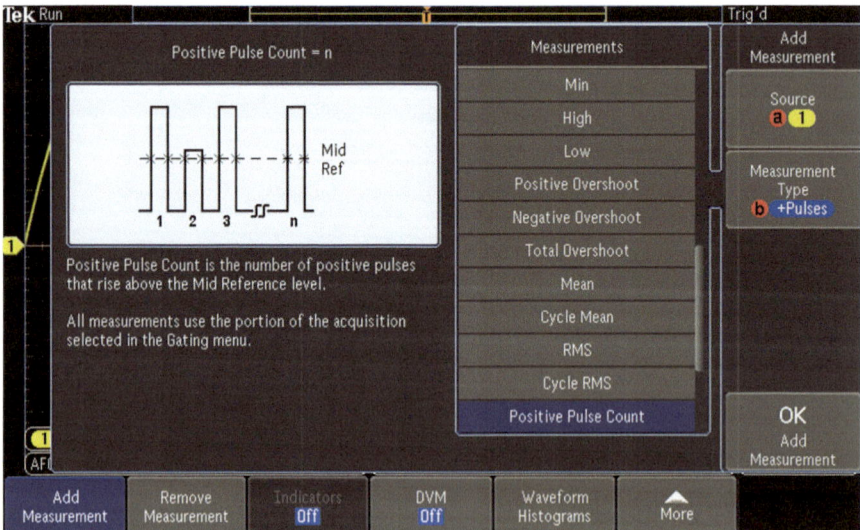

Fig. 4.8 Positive Pulse Counter. (Author's screenshot)

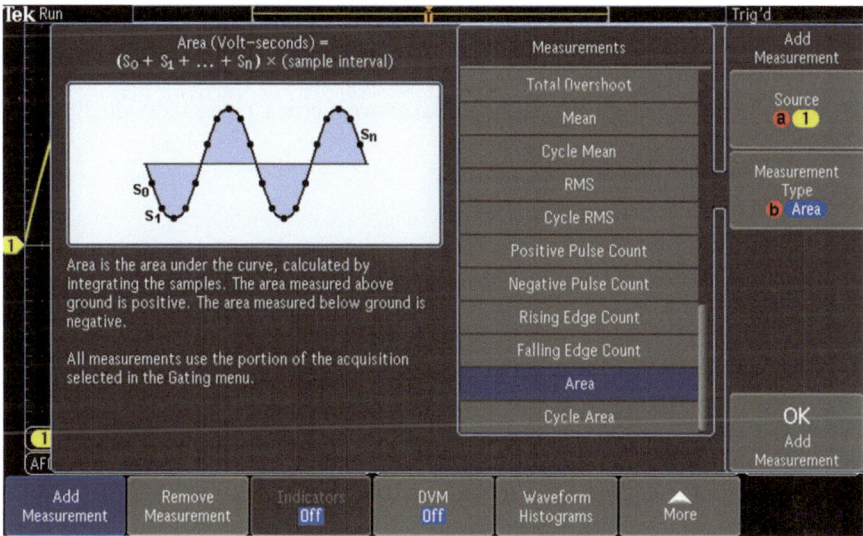

Fig. 4.9 Integrating the samples. (Author's screenshot)

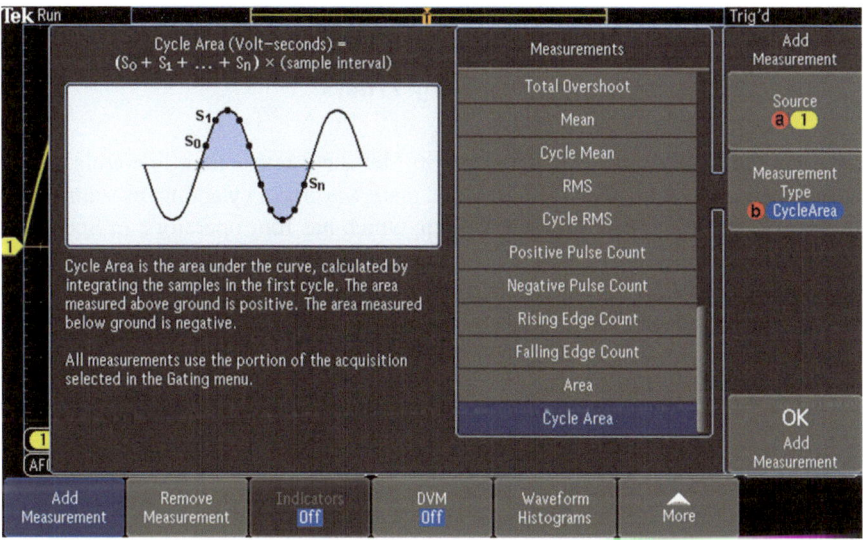

Fig. 4.10 Cycle Area. (Author's screenshot)

Cycle Area, as shown in Fig. 4.10, is the area under the curve, calculated by integrating the samples in the first cycle. The area measure above ground is positive. The area measured below ground is negative.

Remove Measurement deletes any of the above metrics, either individually or all at once.

Digital Voltmeter

A very frequently used oscilloscope mode is Digital Voltmeter (DVM). To access it, first press Default Setup. Then, with AFG on and Math off, press Measure in the Wave Inspector section. Then, press DVM, bringing up the vertical DVM side menu. The DVM default mode is Off. It turns on when one of the side menu selections is chosen. The AFG default waveform (Sine) has a predetermined peak-to-peak voltage, 500.00 mVpp, and frequency, 100.00 kHz. They are displayed in the trace, if you count divisions, and in the measurements shown at the top in large, easy to read figures. Note that RMS (root mean square) for 500.00 mVpp is a little over 163 mV. This is the effective voltage as in heating a resistive element, and it is the value of interest in most applications. RMS is defined as the square root of the arithmetic mean of a set of numbers regardless of the sign.

Adding and Subtracting Frequency Traces

Spectrum Math resembles Dual Waveform Math, but rather than operating in the time domain, it permits the user to create a math waveform by adding or subtracting frequency traces. (Unlike Waveform Math, which has four operators, in Spectrum Math there are only two operators, Add and Subtract. That is because it does not make sense to multiply or divide one frequency by another.)

To access Spectrum Math, begin by defaulting the instrument. Then, press Math (twice) and RF. You will see the orange RF trace below and the red Math trace above, as shown in Fig. 4.11.

The amplitude levels at each frequency are the same. This may be more easily seen by pressing Stop. Then return to Run.

The Spectrum Math menu appears across the bottom of the display. Pressing the soft key corresponding to Spectrum Math generates a side menu, which permits the user to construct the desired Spectrum Math trace.

In addition to the operator, the Spectrum Math side menu allows the user to set the first and second sources, using Multipurpose Knobs a and b respectively. The available choices are RF:N, A, M, and m. Preset Labels are ACK, ADO, ADDR, ANALOG, BIT, CAS, CLK, CLOCK, CLR, COUNT, DATA, DTACK, ENABLE, HALT, INT, IN, IRQ, LATCH, LOAD, NMI, OUT, PIN, RAS, READY, RESET, RX, SEND, SHIFT, STROBE, TX, and WR_.

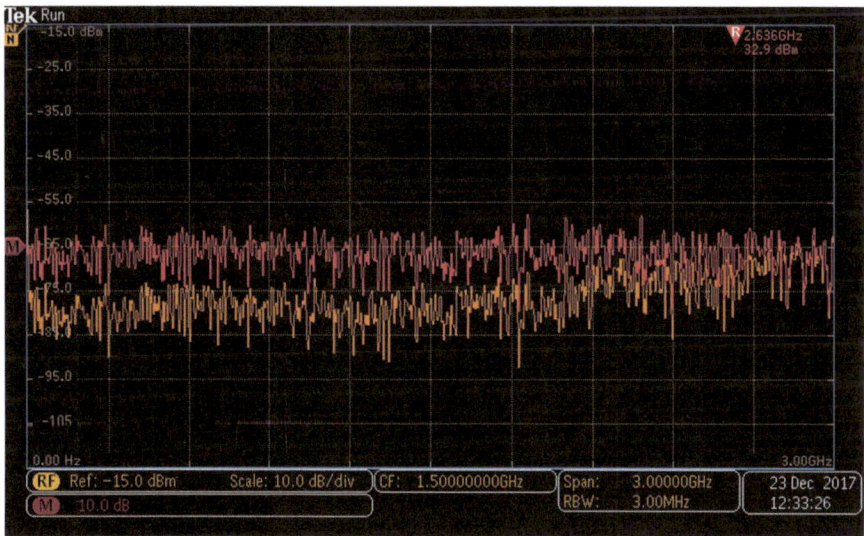

Fig. 4.11 Spectrum Math. (Author's screenshot)

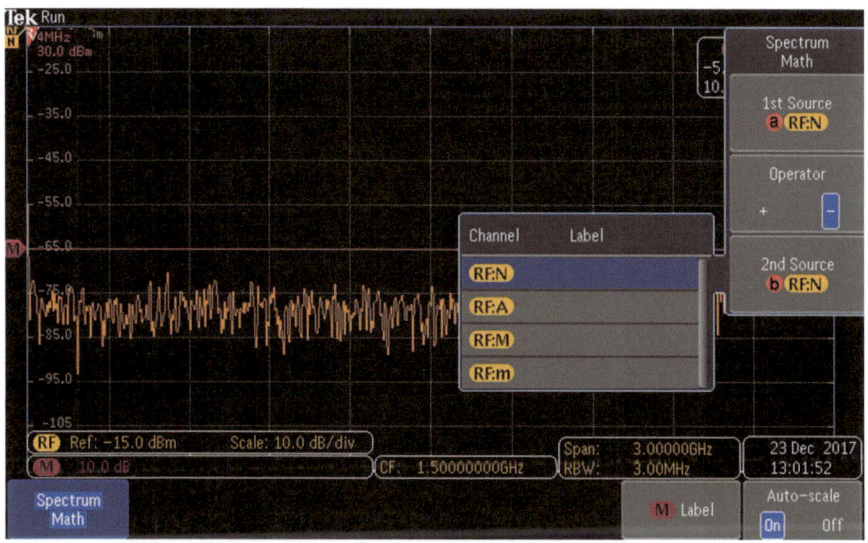

Fig. 4.12 Sources, operators and labels are chosen by turning Multipurpose Knobs and toggling a soft key. (Author's screenshot)

An infinitude of additional labels can be created by using the on-screen keyboard, as shown in Fig. 4.12.

In DVM, we can look at AC Plus DC RMS, DC, AC RMS, and Frequency, as shown in Figs. 4.13, 4.14, 4.15, and 4.16.

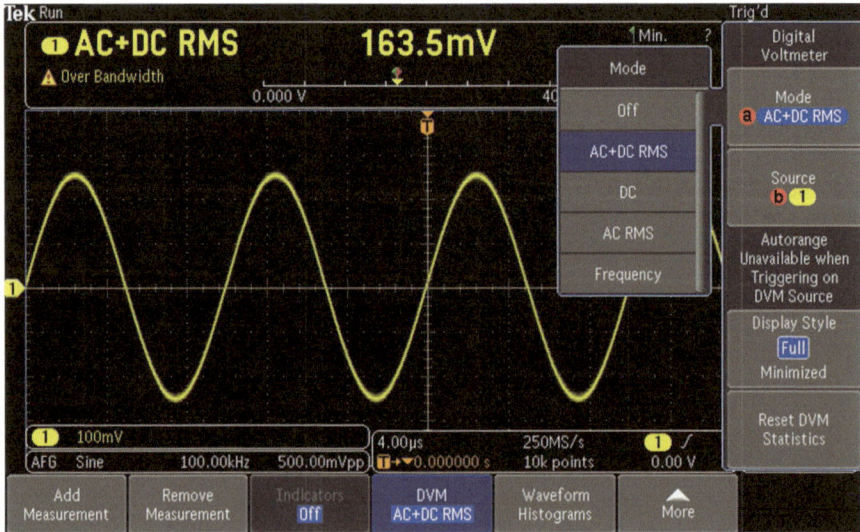

Fig. 4.13 AC plus DC RMS. (Author's screenshot)

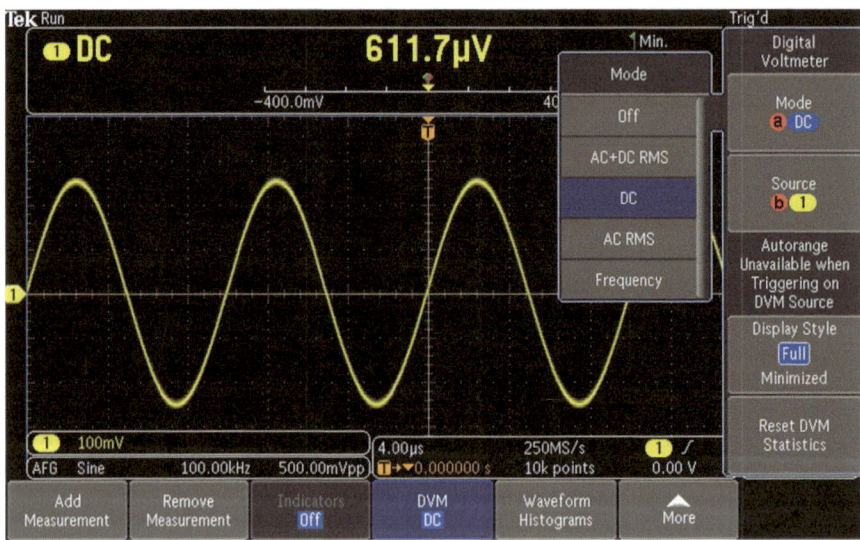

Fig. 4.14 DC. (Author's screenshot)

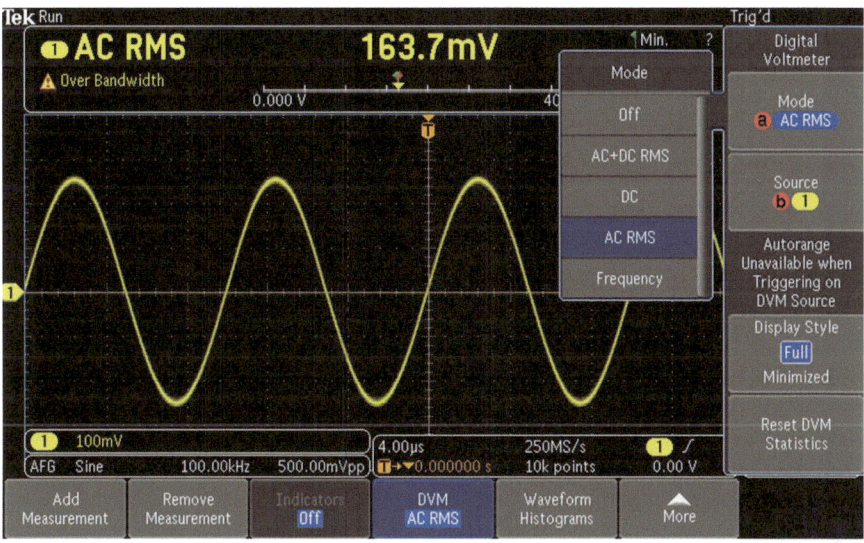

Fig. 4.15 AC RMS. This is useful in quantifying ripple in a DC power supply. (Author's screenshot)

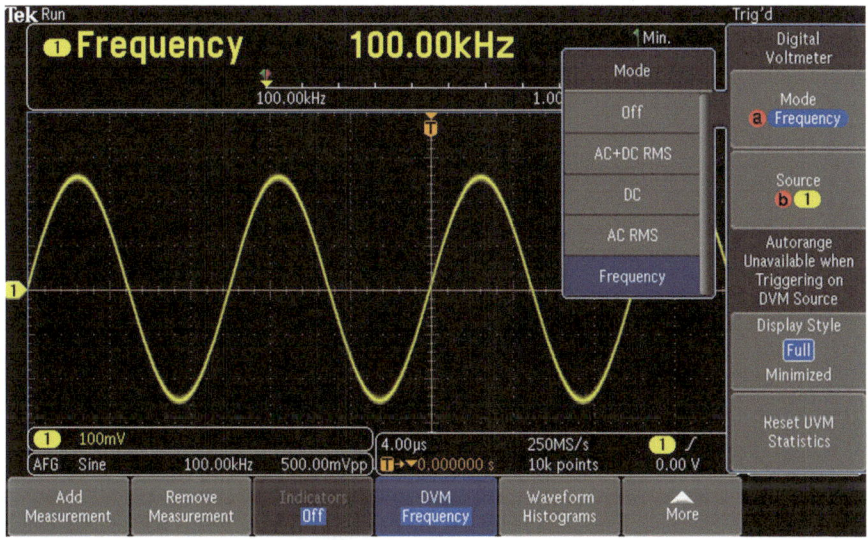

Fig. 4.16 Frequency. (Author's screenshot)

Any of these values can be changed by going back to AFG > Waveform Settings and the DVM readings will reflect these changes.

Waveform Histograms

Besides signal measurements in an oscilloscope, histograms are used in numerous fields such as census data, actuarial statistics, and digital photography, where an image histogram consists of a graphical representation of the tonal distribution. The graph displays the number of pixels for each tonal value. A post-processing technician can use sliders or type numeric values in order to modify hue and intensity among other parameters. The histogram changes to reflect the photo's evolving form. Similarly, in a digital oscilloscope, histograms can be invoked to display graphically continuous variables.

As the user presses front panel buttons, vertical (voltage) and horizontal (time) histograms respond as shown in the instrument's automated measurements.

In the popular imagination histogram implies a historical meaning of the diagram. But in actuality, the word derives from the ancient Greek words ἱστός (histos), which refers to rows of vertical objects such as cornstalks, and γράμμα (gramma), which means an illustration.

Karl Pearson (1857–1936), shown in Fig. 4.17, was an English mathematician and biostatistician who played a key role in the science of statistical analysis. He invented the histogram.

Fig. 4.17 Karl Pearson. (Wikipedia)

The histogram looks like a bar graph, but conventionally a bar graph has space between each bar, while in the histogram the rectangles touch one another. The overall outline is irregular, which makes sense because the values are continuous. As the electrical signals in an oscilloscope display change, the histogram is seen to change form.

To create a histogram, automatically or manually, the data is divided and inserted into bins. Each bin becomes a rectangle, its height proportional to the amount of data in the bin.

Since the bins are all the same width, the height is proportional to the bin's population count. Another arrangement is to have bars of differing widths, where each bin's area represents the number of elements and each bin's height corresponds to the population density.

Because the data is continuous, the number of bins may vary. Wide bins reduce noise introduced by random sampling. Narrow bins promote greater precision and resolution. In addition to possessing different bin widths, histograms can by symmetrical or have various shapes.

To access a histogram, begin by displaying the waveform. With an AFG sine wave displayed, in waveform settings reduce the frequency to 100 Hz. Press Measure, then the Waveform Histograms soft key. In the Waveform Histograms menu, Waveform Histograms are Off. The vertical or horizontal histogram can be displayed by pressing the appropriate soft key. The vertical histogram represents amplitude (Fig. 4.18).

The positive peak of the waveform represents maximum value, and the negative peak is minimum value. Under More, you can toggle between lineal and log modes. For this particular display, lineal is the more revealing. Going to the horizontal

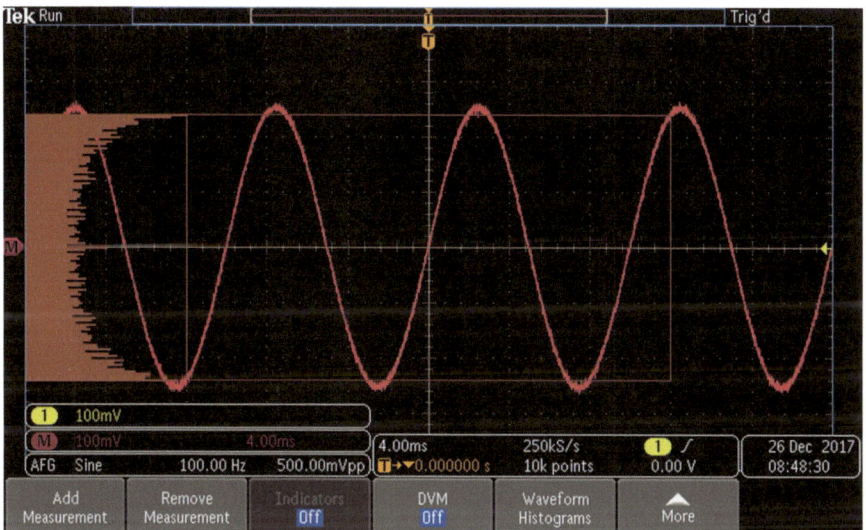

Fig. 4.18 Vertical Histogram of a 100 Hz sine wave. (Author's screenshot)

display, we see that the histogram, as shown in Fig. 4.19, now situated along the time base, correlates not to amplitude but to the rate of change.

A horizontal square wave Histogram still loading data, as shown in Fig. 4.20, has a totally different appearance.

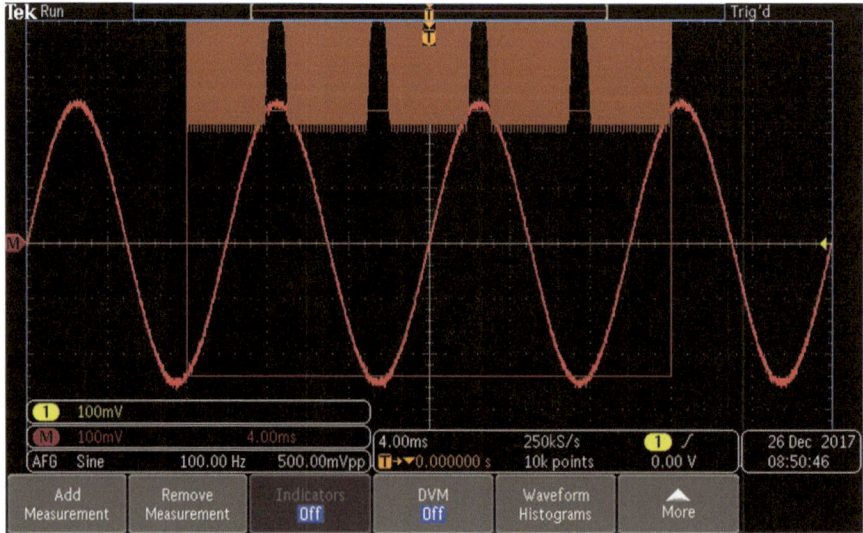

Fig. 4.19   Horizontal Histogram of a 100 Hz sine wave. (Author's screenshot)

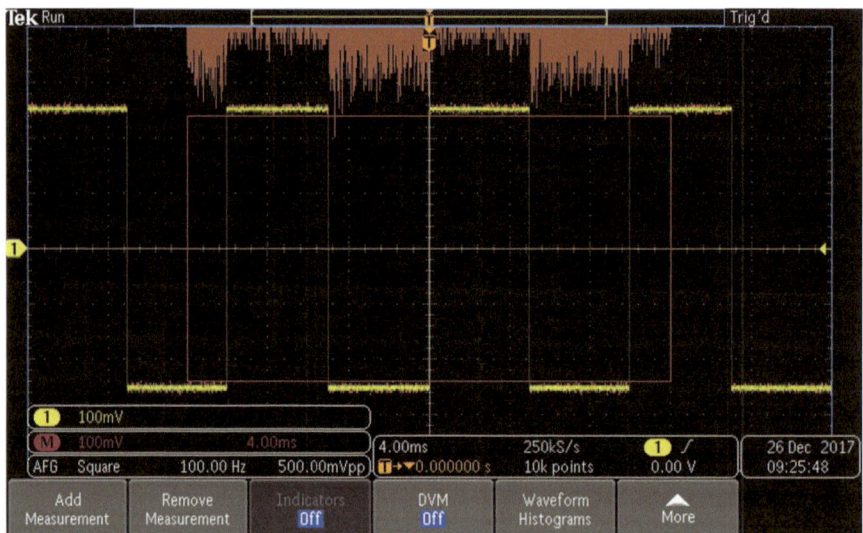

Fig. 4.20   Horizontal Histogram of 100 Hz square wave

Press AFG to enter the waveform menu, and use Multipurpose Knob a to access this interesting waveform, with its near instantaneous rise and fall times. Again change the frequency to 100 Hz. Press Measure and activate the vertical and horizontal displays. The histograms are radically different, with altered amplitudes and rates of change.

Here, in Figs. 4.21 and 4.22 are the vertical and horizontal Noise Histograms:

The vertical histogram exhibits a Gaussian distribution as opposed to the horizontal histogram, which represents a uniform distribution along the time axis. Press Menu Off to see a superimposed rectangle representing the histogram's total area within horizontal and vertical limits. When selected, they may be modified by turning Multipurpose Knobs a and b.

Digital Voltmeter

One of the Measure menu items is DVM, which stands for digital voltmeter. It is a very useful adjunct to the conventional oscilloscope mode because it permits the user to make voltage and frequency measurements of the signal that is being probed without having to fire up a separate instrument.

When the soft key associated with DVM is pressed, the Digital Voltmeter menu appears to the right of the display. The top item is Mode, which is controlled by Multipurpose Knob a. The default setting is Off. Turning Multipurpose Knob a, the

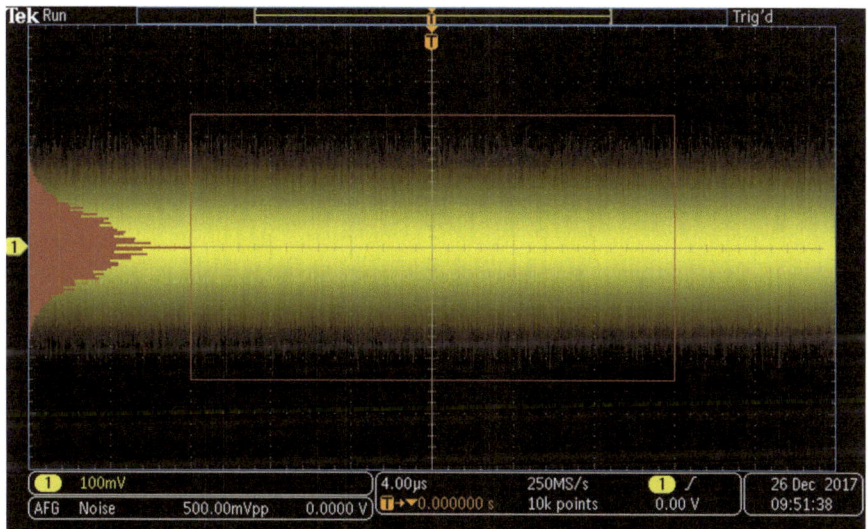

Fig. 4.21 Vertical histogram of noise signal. Frequency is not applicable for noise. (Author's screenshot)

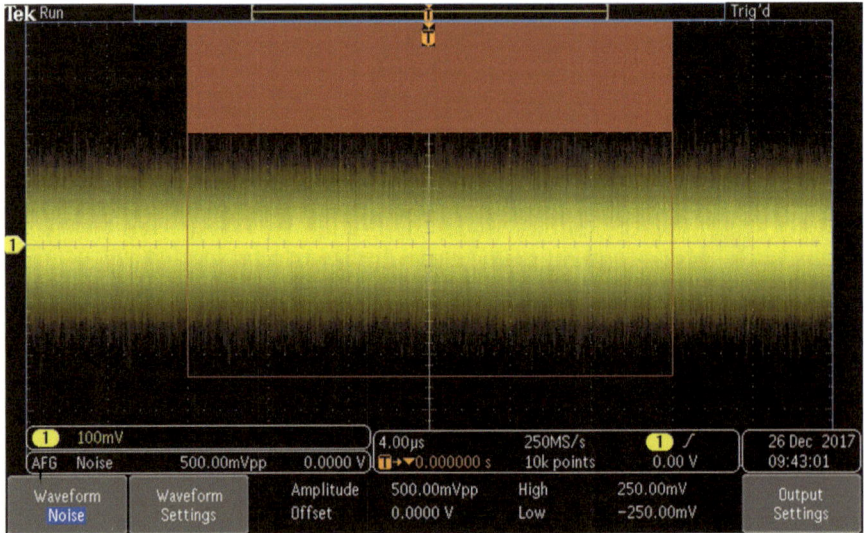

Fig. 4.22 Horizontal histogram of noise signal. (Author's screenshot)

settings other than Off are AC + DC RMS (Root Mean Square), DC, AC RMS, and Frequency. Each of these has a definite purpose, some going beyond the obvious.

AC + DC RMS is commonly used because it provides a lot of information about any given signal. When the positive and negative peaks are equidistant from a zero-point that is referenced to ground, we say that the DC component is zero. That is a bit of a misnomer since the signal is moving through that point and never stationary. That being said, AC + DC is a useful concept and we may say that there is a "DC component" provided the term is in quotes.

Everyone, at least those in the Test and Measurement business, knows that Root Mean Square refers to how AC voltage is measured, and that it is some fraction of peak-to-peak voltage. Actually the concept of root mean square arises in many disciplines, everything from actuarial statistics to the physics of gas molecules, even finding its way onto the floor of the gambling casino.

In an electrical circuit, the RMS value of an alternating current is equal to the amount of direct current that would cause an equal amount of heat to be dissipated passing through a resistive load. As it happens, the RMS value of a continuous waveform is the square root of the arithmetic mean of the squares of the values or the square of the function that defines the continuous waveform. Suffice to say that the RMS voltage is the metric of interest and is what is generally measured by a voltmeter. There are situations where our interests lie elsewhere, and the Measure modes then become relevant. The DC mode is useful for checking the output of a battery or DC power supply. When the DC voltage of the sine wave output of our internal AFG is measured, the DC reading is found to be quite low, in the microvolt range.

AC RMS is very useful in looking at the output and evaluating an AC power supply. Electrolytic filter capacitors are prone to loss of capacitance, and this will manifest as an unacceptable amount of ripple, which can be readily seen in the AC RMS mode of the DVM. The question then becomes how much ripple is unacceptable. The answer varies depending on the size and type of filter network and the intended application. The way to go is to monitor the output ripple in a known good unit and keep records to see if there is an increase.

Finally the Frequency mode is used to ascertain the frequency of an unknown signal and monitor its stability. The DVM mode in the Measure function is frequently used, and it is very user-friendly.

Statistics

Another Measure submenu is Statistics. To access it, first press Measure. Then, in the Measure menu that appears across the bottom, press More so that a pop-up menu enables additional menu selections. If necessary, toggle More until Statistics is highlighted, revealing the Statistics side menu. The top soft key toggles it On.

To see how Statistics works, we need a signal at one of the inputs. So, we connect the internal AFG output on the rear panel via a BNC cable to the Channel One analog input. Pressing AFG (twice) activates the defaults. Channel One turns on, and the AFG output is a sine wave with a 500.00 mVpp value at 100.00 kHz.

Going back to Measure and turning on Statistics, we see that the number of mean and standard deviation samples can be set by turning Multipurpose Knob a. The choices range from two to 1000. For now, we leave the number of samples at the default, 32. Next, to initiate the operation, press Add Measurement. Multipurpose Knob a selects Source, which is defaulted at Channel One.

Multipurpose Knob b selects the Measurement Type. There are 31 categories, ranging from Snapshot All to Cycle Area, as detailed earlier in this chapter.

For now, we'll look at Frequency. Using Multipurpose Knob b, highlight Frequency and press OK Add Measurement, as shown in Fig. 4.23.

When AFG is pressed, the sine wave is shown, as in Fig. 4.24, with Statistics including Value, Mean, Minimum, Maximum, and Standard Deviation.

Now we'll look at a different waveform and parameter. Press Waveform and using Multipurpose Knob a, select Square Wave. Leaving Source set at Channel One, use Multipurpose Knob b to shift the Measurement Type to Fall Time, as shown in Fig. 4.25.

Press OK Add Measurement to activate Fall Time and press AFG to see the Fall Time Statistics for the Square Wave, as shown in Fig. 4.26

Now, for perspective, we'll look at another waveform. Return to AFG and using Multipurpose Knob a choose Noise. Returning to Measure, select as the parameter Amplitude, as shown in Fig. 4.27.

Figure 4.28 shows amplitude for the Noise signal. Notice that the previous Statistics, Frequency for the Sine Wave, and Fall Time for the Square Wave are retained in the display.

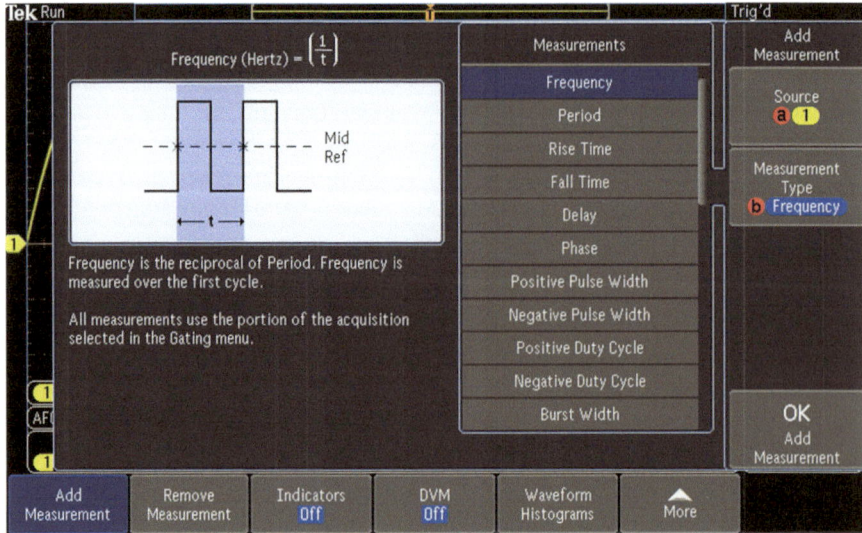

Fig. 4.23 Frequency is selected and the measurement is added. Notice the relevant equation, a generic diagram illustrating the concept of the selected parameter and the description. (Author's screenshot)

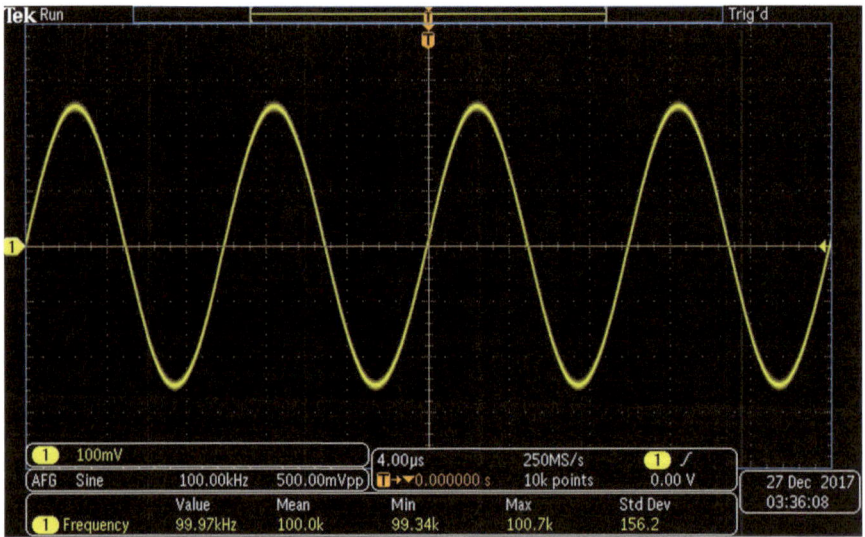

Fig. 4.24 Sine Wave is shown with Statistics for a 100.00 kHz (nominal) trace. (Author's screenshot)

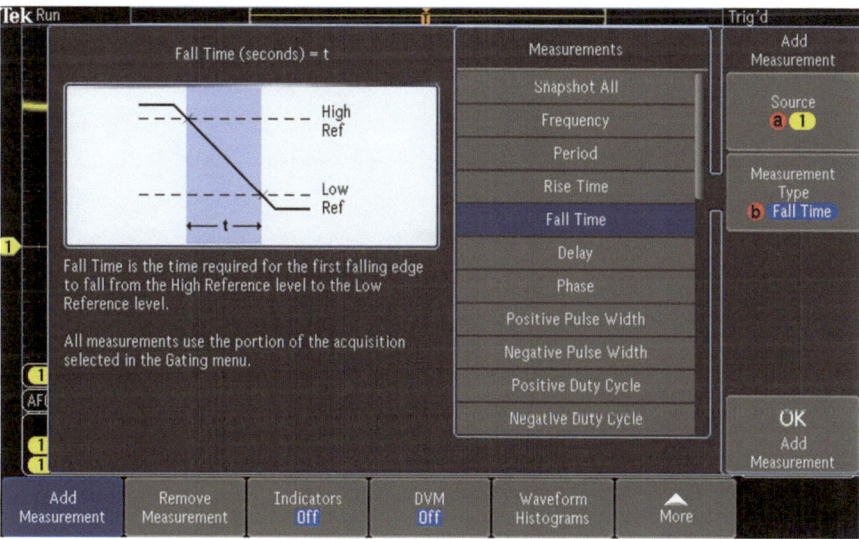

Fig. 4.25 With Measurement Type set at Fall Time, the relevant equation, generic waveform, and definition are shown. (Author's screenshot)

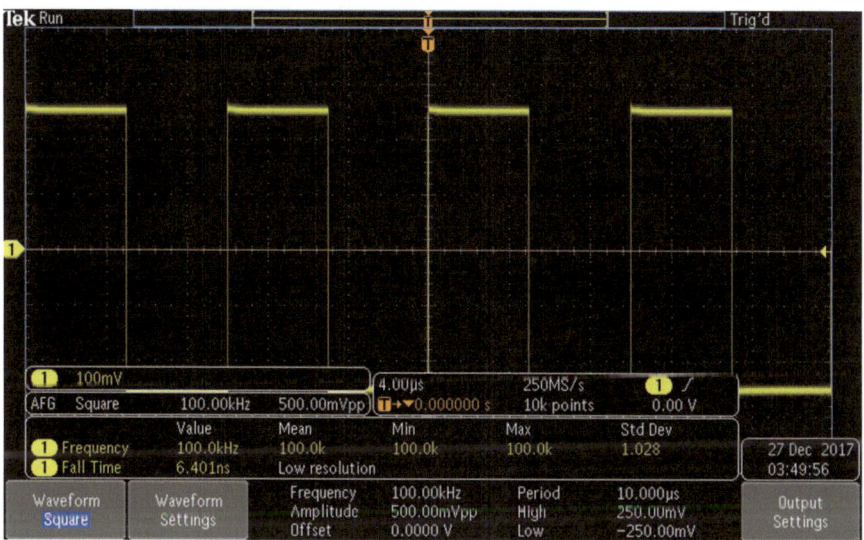

Fig. 4.26 Fall Time for Square Wave is 6.40 ns. Notice that the previous statistic, Sine Wave Frequency, is retained in the display. (Author's screenshot)

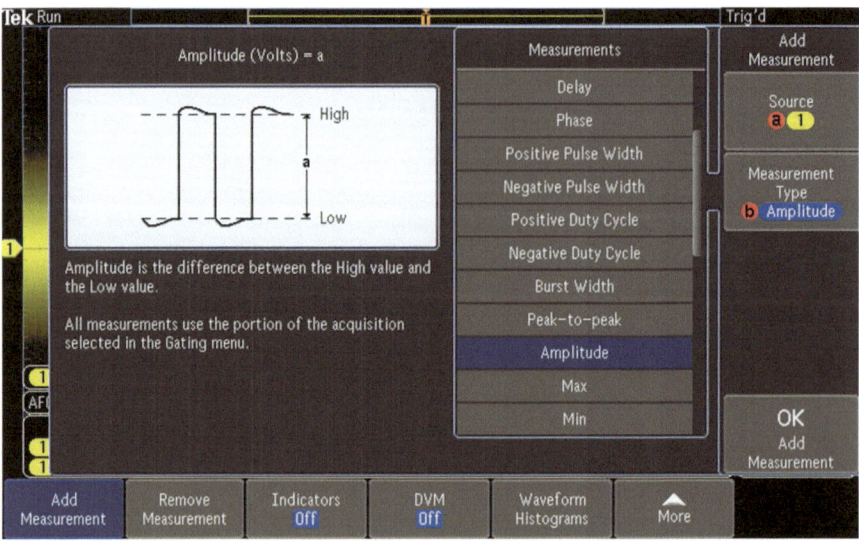

Fig. 4.27 Equation, generic waveform diagram and definition are shown for Amplitude. (Author's screenshot)

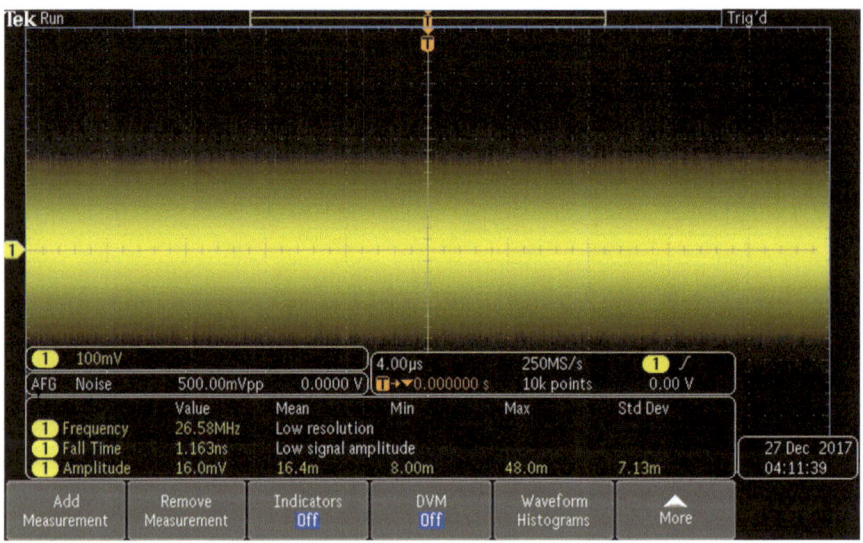

Fig. 4.28 Amplitude for a Noise signal. (Author's screenshot)

Gating

Another subtopic that is relevant to Measure is Gating. It applies the measuring operation to a portion of the waveform that is intentionally limited by the user. To initiate Gating, press Measure. Then, in the Measure menu at the bottom of the screen, press More as necessary to highlight Gating. The Vertical Gating submenu appears at the right of the screen. With the internal (or an external) AFG applying a sine wave to channel One, the Gating menu selections are Off between Cursors and Bring Cursors On Screen. The purpose of these options is to permit the user to position the gates as desired to constrain measurements to limited time intervals, as shown in Fig. 4.29.

Gating is a valuable tool when you are probing a bus and want to measure an amplitude only when a given voltage is present at another location. Likewise, an irrelevant high-amplitude spike can be eliminated. Another application is when measuring a voltage that varies between two states, only one of which is currently of interest.

Reference Levels

Continuing in the Measure Menu in More, the next selection following Gating is Reference Levels. These are vertical lines that intersect the X-axis and intersect waveforms that are plotted against the Cartesian coordinates comprising the oscilloscope display.

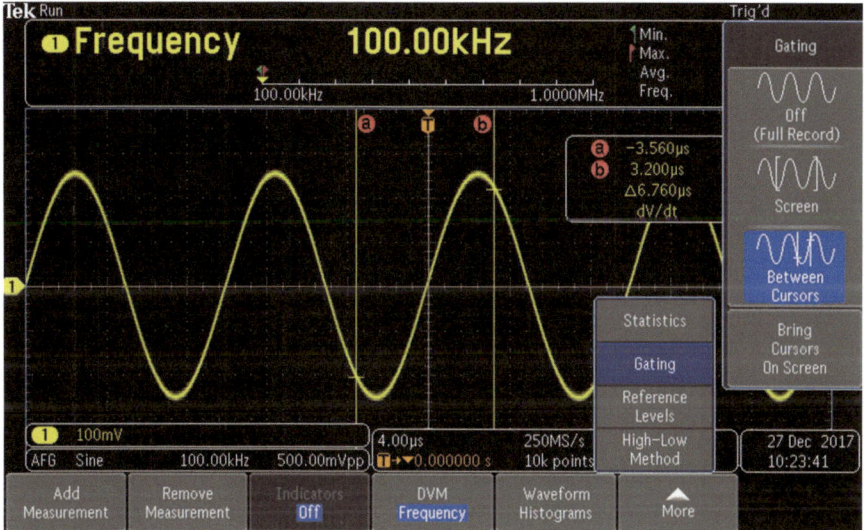

Fig. 4.29 Gating applied to a 100.00 kHz sine wave. (Author's screenshot)

They are placed by the user. In the vertical Reference Levels menu, levels can be set in either percentages or in units by toggling the top soft key. Then, High, Mid, and Low References can be set using Multipurpose Knobs or the number pad. Most typically, the lowest soft key is pressed to set Reference Levels to default amounts.

Next, press Add Measurement to select the desired parameter, using Multipurpose Knob a. Then press OK Add Measurement. Measurements may require one, two, or three Reference Levels. In Figs. 4.30 and 4.31, Rise Time in a Ramp Waveform has been chosen, with default settings. Press the Reference button, below Math, to see the waveform with reference lines and rise time values, with Mean, Minimum, Maximum, and Standard Deviation statistics.

The final menu selection is High-Low Method. Either of two methods for denoting high level and low level can be selected in the High-Low menu (under Measure>More). The Histogram is better for pulses and Min-Max works well for all other waveforms.

Automatic Measurements in Time and Frequency Domains

Automatic measurements in an oscilloscope are in contrast to manual measurements, which are accomplished by placing horizontal and vertical cursors within the display so that they delineate the portions of a waveform that is to be measured. We'll see how automatic measurements are implemented in both time and frequency domains.

To set up an oscilloscope for doing a time domain automatic measurement, power up the instrument and bring it to default status so that irrelevant settings from

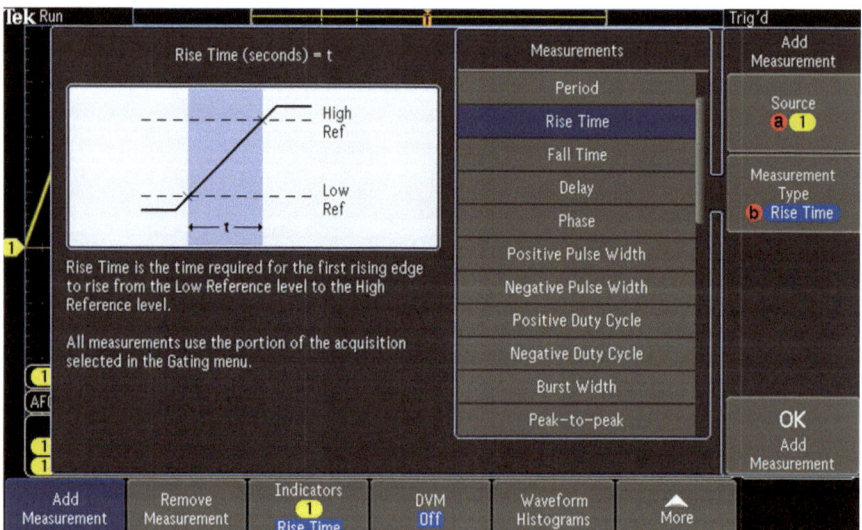

Fig. 4.30 Rise Time is set as Measurement Type. (Author's screenshot)

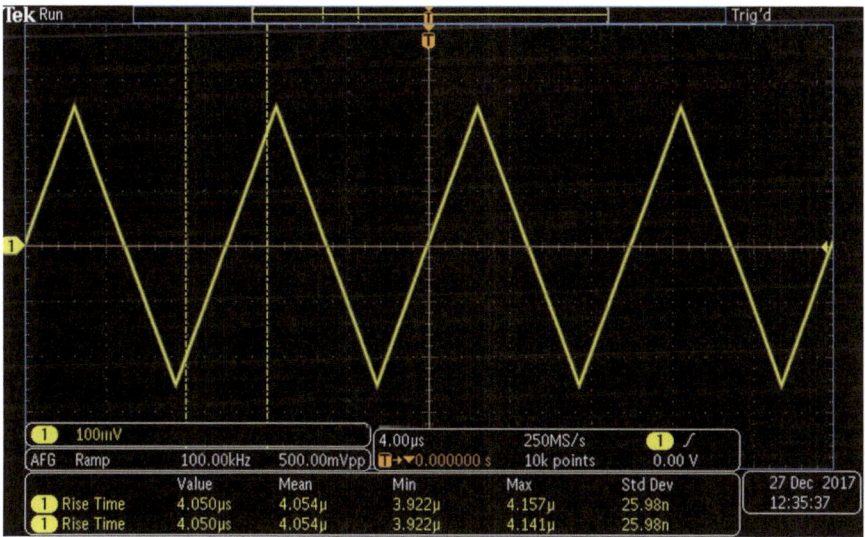

Fig. 4.31 Ramp Wave shown with Reference Levels and Values displayed. (Author's screenshot)

a previous work session do not interfere. Make sure that the machine is in the time domain by pressing the button corresponding to the analog channel to be used as input.

Press Measure in the Wave Inspector section. In the Measure menu at the bottom of the display, press the soft key associated with Add Measurement. The vertical Add Measurement menu appears at the right of the display. The top menu selection, Source, displays the currently active channel, which can be changed by turning Multipurpose Knob a.

The second menu selection permits the user to choose the measurement type. The default mode is Snapshot All Measurements. To illustrate, let us suppose we are interested in Peak-to-peak. Use Multipurpose Knob b to scroll through the list of measurements, highlighting the desired measurement as shown in Fig. 4.32.

One or all measurements can be removed by pressing the Remove Measurement soft key and completing the action in the submenu.

So far, the automatic measurement has been initiated but the action has not been completed. To conclude the operation, press OK Add Measurement. Then, to see the list of completed measurements, press the active channel button. Press AFG to see the original sine wave in the same display, as shown in Fig. 4.33.

In displaying automatic measurements, if a triangle icon with an exclamation point appears, there is a vertical clipping situation. Turn the vertical scale and position knobs so that the entire waveform appears in the display. If a low-resolution message appears, increase the record length of the acquisition so that the oscilloscope has more points from which to calculate the measurement.

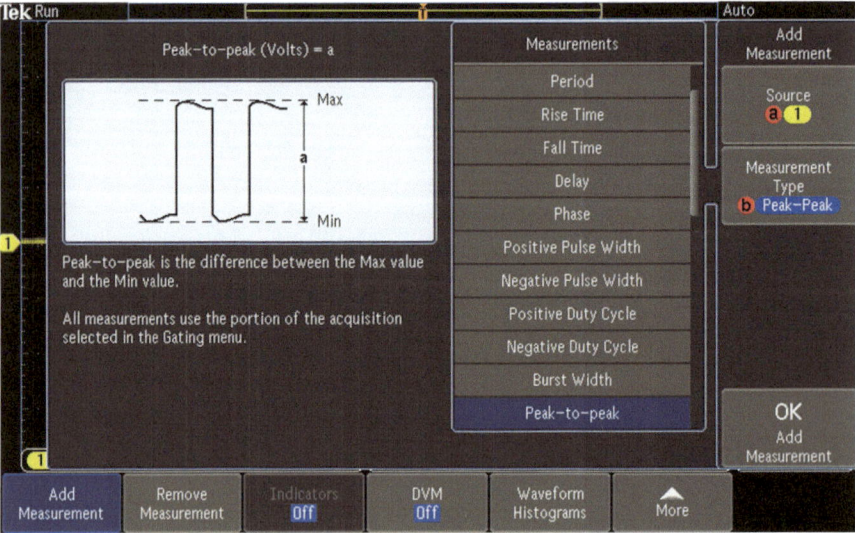

Fig. 4.32 Peak-to-peak is selected from the list of measurement types. At top left is the applicable equation. The generic waveform diagram clarifies that the peak-to-peak voltage is between reference levels at the absolute Maximum and Minimum values without regard to other peaks that may occur in a complex waveform. (Author's screenshot)

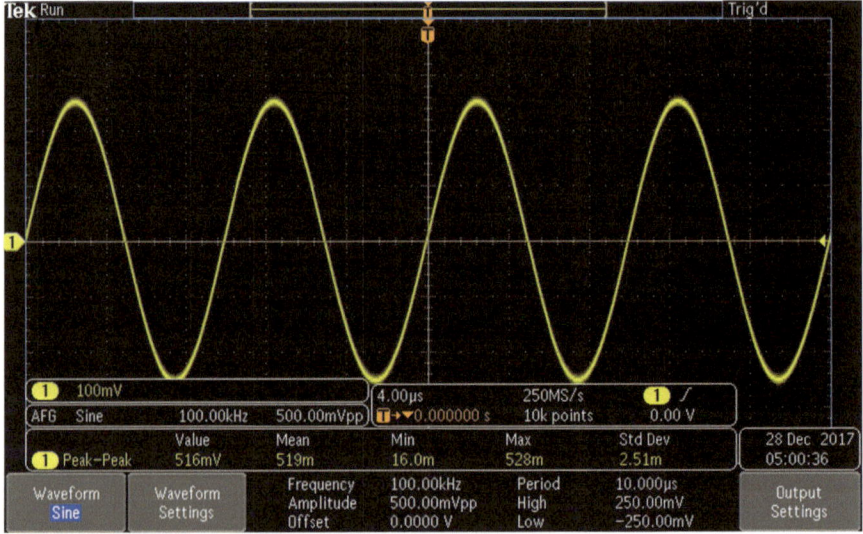

Fig. 4.33 A sine wave with completed automatic measurements. The measurement bar at the bottom of the display lists measurements with source, measurement type, value, mean, minimum, maximum and standard deviation. (Author's screenshot)

Automatic Measurements in the Frequency Domain

The procedure for taking automatic measurements in the frequency domain is similar to the time domain procedure, but due to the different X- and Y-axis definitions, the menu selections are different. To initiate automatic measurements in the frequency domain, after default setup, press RF. Shift the BNC cable that comes from the internal AFG to the RF input, using an RF adapter.

The RF display consists of a prominent noise floor with a single spike for the sine wave fundamental. There are no harmonics. The signal may be difficult to see because it coincides with a vertical grid line at the left edge of the display. To move the sine wave fundamental to the center of the screen, press Freq/Span, bringing up the vertical Frequency and Span menu. The top menu selection is Center Frequency, which currently stands at the default 1.50 GHz. Using the number pad and units soft key, change the center frequency to 100.00 kHz to match the current AFG sine wave frequency. Notice now that the fundamental has migrated to the center of the screen, where it is much easier to see, as shown in Fig. 4.34.

To see how harmonics appear in the frequency domain, we'll press AFG and switch to square wave, a signal that is very rich in harmonics due to the extremely fast rise and fall times, which are essentially high-frequency wave components. No harmonics are visible in the display until the Span is increased to 12.0 MHz. Now the Start and Stop frequencies automatically fall in line at −5.90 MHz and 6.10 MHz, respectively. Many harmonics stand high above the noise floor, as shown in Fig. 4.35.

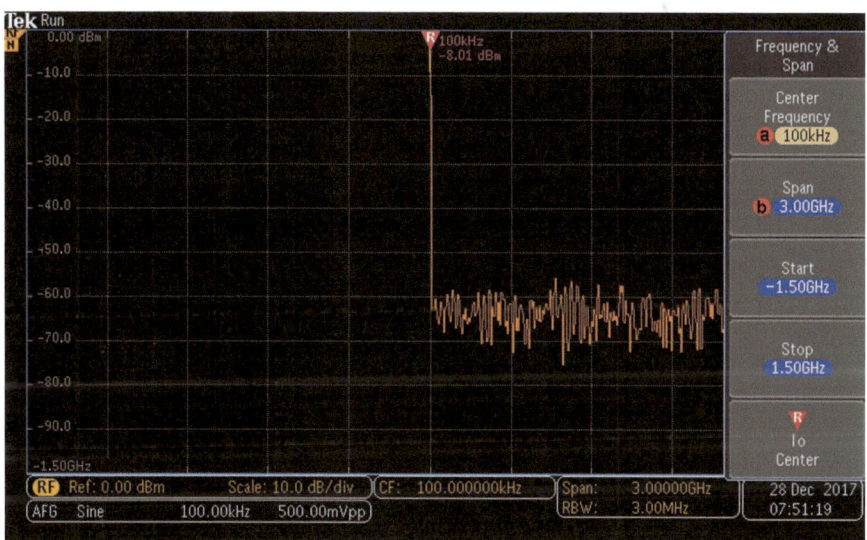

Fig. 4.34 AFG and frequency domain center frequency coincide at 100.00 kHz, so the fundamental is at the center of the screen. There are no harmonics because in a sine wave all the power is at the fundamental frequency. (Author's screenshot)

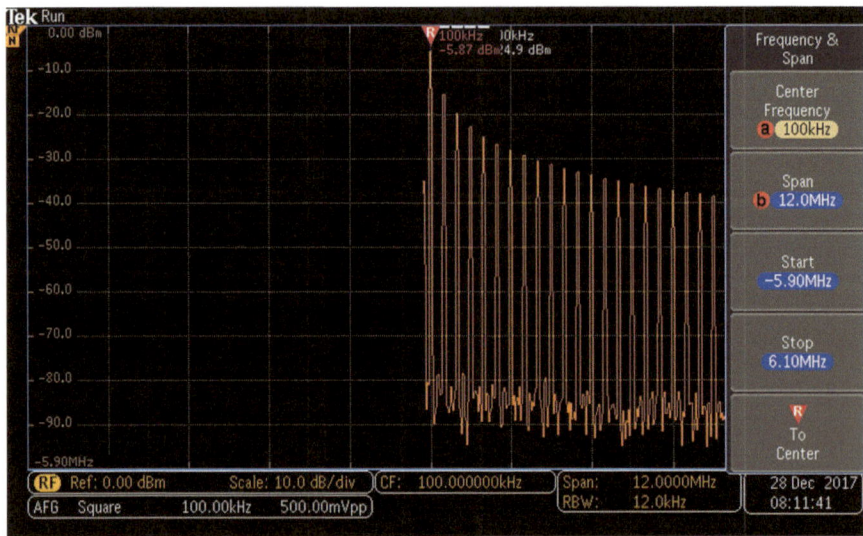

Fig. 4.35 A square wave with harmonics beginning at about 10 dBm below the fundamental and declining in power with spectral distance from it. (Author's screenshot)

This is the trace we will use to demonstrate automatic measurements in the frequency domain. With the 100.00 kHz square wave displayed in the frequency domain, press Measure. The only menu selection in the horizontal menu below the display is Select Measurement, currently set at None. Pressing the adjacent soft key, the vertical Select Measurement menu appears at the right of the display. Menu selections are:

- Channel Power, which is the total power within the bandwidth, defined by the channel width
- Adjacent Channel Power Ratio, which is the main channel and the ratio of channel power to main power, for the upper and lower halves of each adjacent channel
- Occupied bandwidth, which is the bandwidth that contains the specified percent of power within the analysis bandwidth

  In making measurements in the frequency domain, relevant values are assured by onscreen help explaining the purpose of each measurement. Configuration information is shown in the horizontal menu. Pressing configure sets the measurement parameters in a side menu and the instrument sets the span. With RF Measurements On, Auto Detection determines the frequency domain display using Average Detection. Channel Power, Adjacent Channel Power Ratio and Occupied Bandwidth are shown in Figs. 4.36, 4.37 and 4.38. They are configured in Figs. 4.39, 4.40, and 4.41.

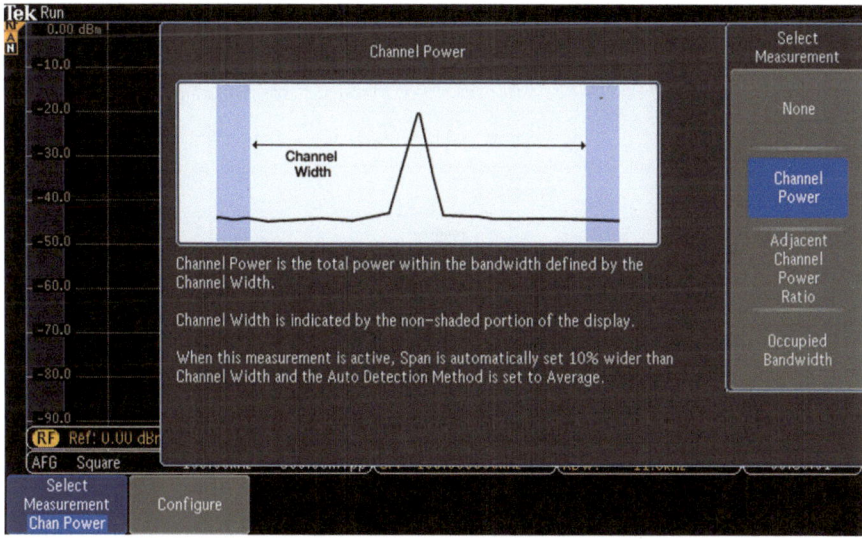

Fig. 4.36 Channel Power information. (Author's screenshot)

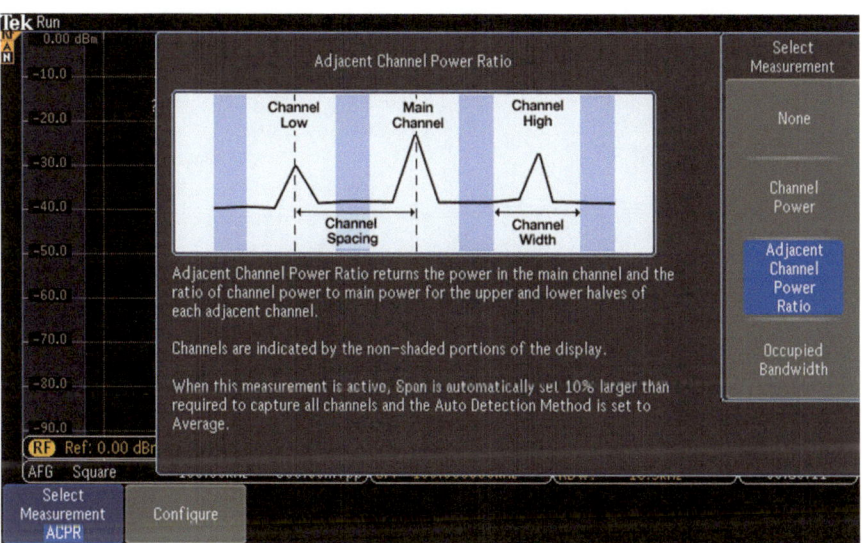

Fig. 4.37 Adjacent Power Ratio information. (Author's screenshot)

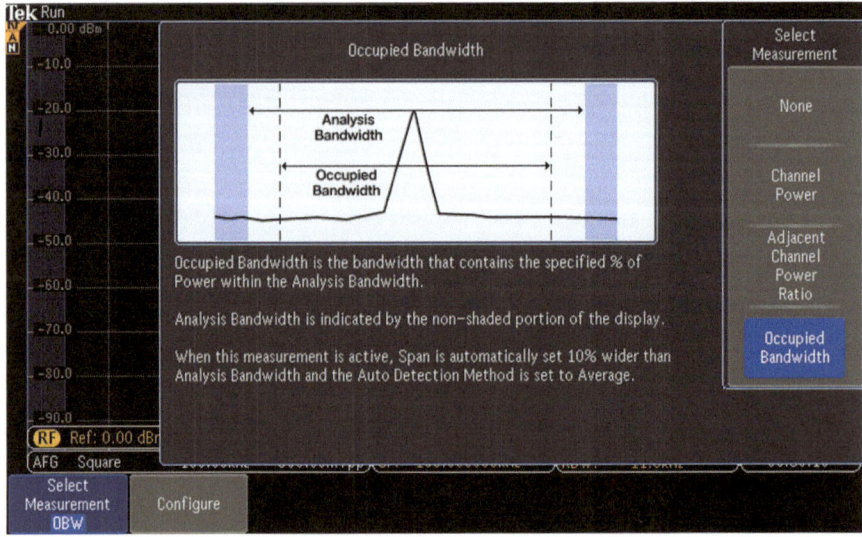

Fig. 4.38 Occupied Bandwidth information. (Author's screenshot)

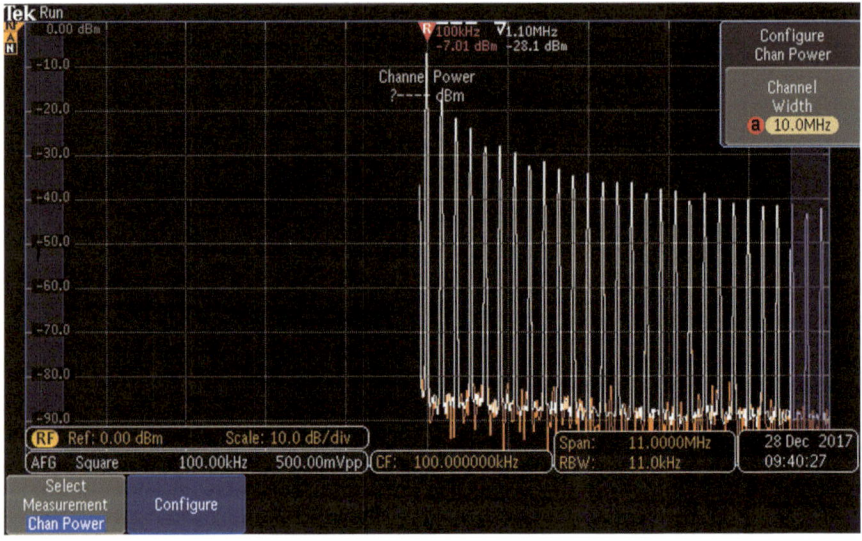

Fig. 4.39 Channel Power configured. (Author's screenshot)

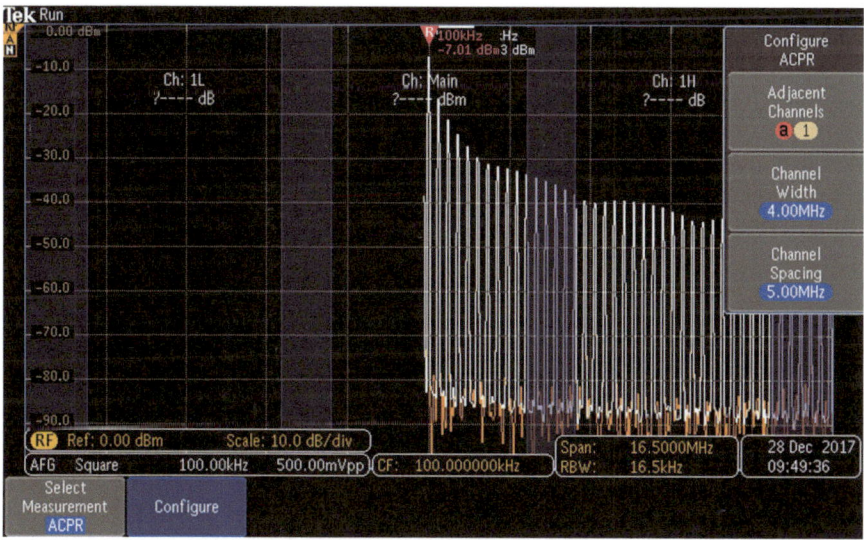

Fig. 4.40 Adjacent Channel Power configured. (Author's screenshot)

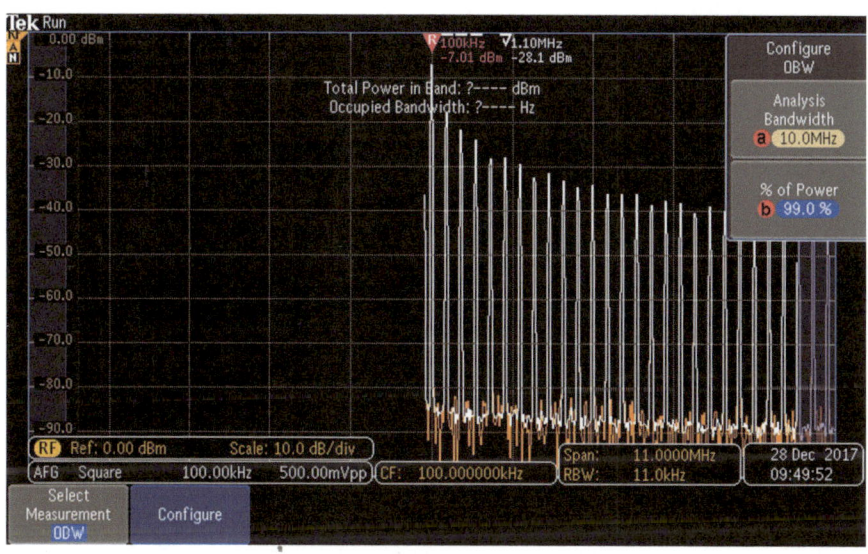

Fig. 4.41 Occupied Bandwidth configured. (Author's screenshot)

Manual Measurements Using Cursors

Cursors are very useful tools that permit oscilloscope users to bracket a portion of an analog or digital waveform for the purpose of measuring the signal displayed. (We'll discuss manual measurements of digital data in Chap. 6. For now, we're focusing on analog analysis.) Manual measurements work only in the time domain. If the instrument is in the frequency domain and you press the cursor button, it won't light and in fact there is no such functionality.

So now we'll swing the AFG output back from RF to analog input Channel One, and press the associated button if it is not currently lighted. As another preliminary, press AFG to display the default sine wave.

To enter the world of manual measurements, press the Cursor button, adjacent to Multipurpose Knob a. This is quite a high-tech button because in addition to a green light, which indicates it is on, a second push of the button turns it off, and if you press and hold, the horizontal Cursors menu appears across the bottom of the screen.

Cursors are horizontal and/or vertical lines that the user can move throughout the display to facilitate manual measurements on displayed data. The first menu selection in the horizontal cursor menu allows the user to choose between Waveform, which is default when Cursors is first turned on, and Screen. In the Waveform mode, only the vertical cursors are shown in the display. They can be moved, the left cursor by means of Multipurpose Knob a and the right cursor by means of Multipurpose Knob b.

The fourth cursor menu selection sets whether or not the cursors are linked. When they are linked, Multipurpose Knob a moves them together, so they always remain the same distance apart. Multipurpose Knob b moves only the right cursor. That way, the distance between them can first be set and then they can be moved together.

Whenever cursors are active as determined by the cursors button, the cursors readout appears as a rectangular box with rounded corners in the upper right corner of the display. The cursors readout contains metrics relative to the onscreen cursors as currently placed across a waveform. The vertical cursors intersect the X-axis and accordingly in the left column the time interval between them is shown. When Multipurpose Knob a is turned, the Delta (change) value remains constant, but when Multipurpose Knob b is turned, the Delta value varies according to the distance between the two vertical cursors.

In the right column, amplitude in mV is shown. As the vertical cursors are moved, the points where they intersect the waveform (as indicated by short horizontal line segments) change and so the amplitude values in the cursor readout vary.

In the first menu selection, Waveform or Screen can be chosen. So far, we have been in the Waveform mode, but now we'll switch to Screen. A second pair of cursors appears. These cursors are horizontal, and they measure distance relative to the Y-axis. Notice that these horizontal cursors consist of dashed rather than solid lines. This means two things:

- They cannot be moved by means of the multipurpose knobs or in any other way.

- The values they enclose are not shown in the cursor readout.

This pair of cursors is shown on the screen, but it is not active. The third menu selection, Bars, is used to activate a pair of cursors represented by dashed lines, always at the expense of the other pair. In other words, only one pair of cursors can be active at a given time, and the Bars menu selection toggles between them. The horizontal cursors move relative to the Y-axis and measure amplitude as shown in the Cursor Readout.

The last menu selection, Units, permits the user to determine the vertical bar units. Seconds is default, but Hz (1/s), ratio, a percentage based on the cursor positions as 100 percent, or phase, based on the cursor positions as 360 degrees, can also be used. The horizontal cursor units can also be changed by the user, to ratio in percent, based on the cursor positions as 100 percent.

Cursors also work in XY mode, when Lissajous patterns are displayed. When XY mode is on, cursors appear to the right of the lower (XY) portion of the screen. They include rectangular, polar, product, and ratio readouts. The oscilloscope displays vertical cursors in the upper (YT) portion of the screen.

We have seen how cursors are used to perform manual measurements. The parameters shown are not nearly as extensive as in automatic measurements, but when only simple time and amplitude measurements are required, cursors are quick and easy, so manual measurements are often the method of choice.

Dual Waveform Math

Returning to the Math button, the horizontal Math menu appears below the display. The first menu selection on the left is Dual Waveform Math. If the associated soft key is pressed, the vertical Dual Waveform Math menu appears at the right side of the display. The top menu selection is First Source, and it is set by turning Multipurpose Knob a. The bottom menu selection is Second Source, set by Multipurpose knob b. In both instances, the alternatives are Channels One through Four and Reference Waveforms R1 through R4. These are the two signals that are the inputs for Dual Waveform Math.

To demonstrate this interesting oscilloscope capability, we'll choose for the two sources Channels One and Two. The AFG is connected via BNC cable from AFG Out on the back panel to the Channel One analog input on the front panel. A 10:1 passive probe is connected to the Channel Two input. This probe is connected to a nine-volt battery, probe tip to the positive terminal and ground return lead to the negative terminal. (Incorrect polarity causes strange results in Dual Waveform Math.) Make sure both channels and the AFG are active. In the display we see the AFG default sine wave and the flat horizontal trace that corresponds to the nine-volt DC power source. (You may have to press Autoset, depending on the most recent AFG waveform settings. Also, you may have to go back to AFG > Waveform Settings and scroll to Sine Wave.

Press the Math button once more and return to Dual Waveform Math. Press the adjacent soft key so that the vertical Dual Waveform Math menu appears at the right of the display. Press the second soft key repeatedly to toggle through the four operators, Add, Subtract, Multiply, and Divide. Each of these operations is displayed in turn, as shown in Figs. 4.42, 4.43, 4.44, 4.45, and 4.46.

In all cases, the arithmetic operations are performed on an instantaneous, point-by-point basis. Incidentally, if you remove the probe from the analog Channel Two input and in Operators, toggle to Divide, you get an unusual display as shown in Fig. 4.46. That is because you are attempting to divide by zero, which does not have a unique output.

It should be noted that measurements can be taken on Math waveforms as on channel waveforms. Math waveforms are scaled and positioned in the same manner as the Dual Waveform Sources. When source waveform controls are adjusted, the math waveforms also respond.

Fourier Revisited

The second menu selection in the Math menu, following Dual Waveform Math, is FFT. These letters, which stand for Fast Fourier Transform, when you press the relevant soft key open a world of oscilloscope functionality. The instrument instantly becomes a spectrum analyzer, and as such it displays in the frequency domain any signal that is applied at an analog input. The signal is not changed, but the way in which it is displayed is totally different from its time domain version. The reason is that the X- and Y-axes are defined differently.

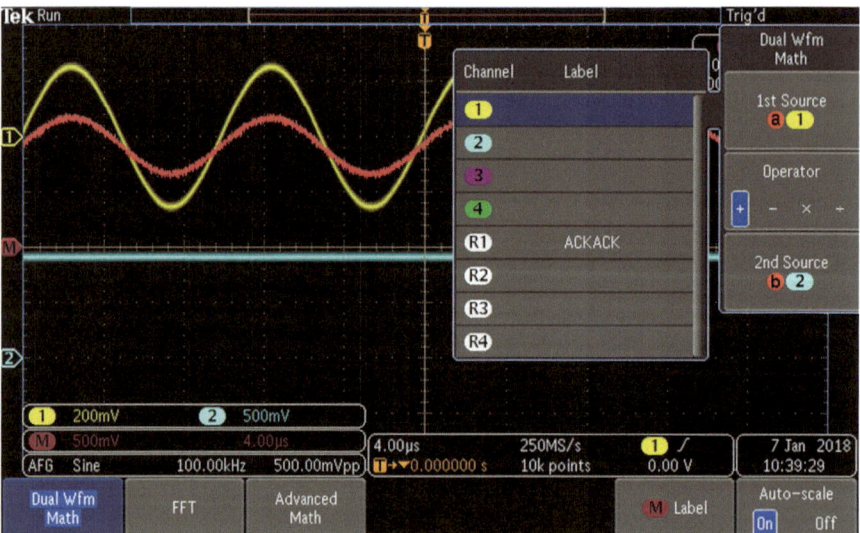

Fig. 4.42 Add. (Author's screenshot)

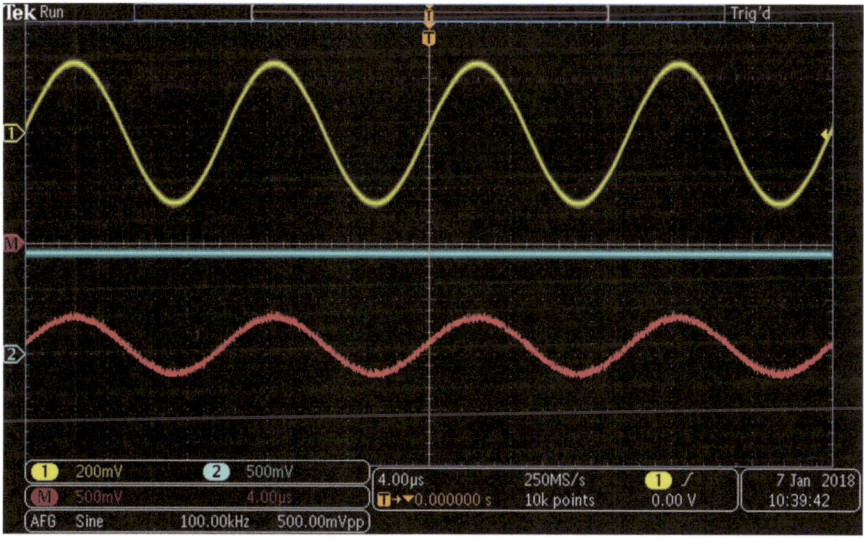

Fig. 4.43 Subtract. (Author's screenshot)

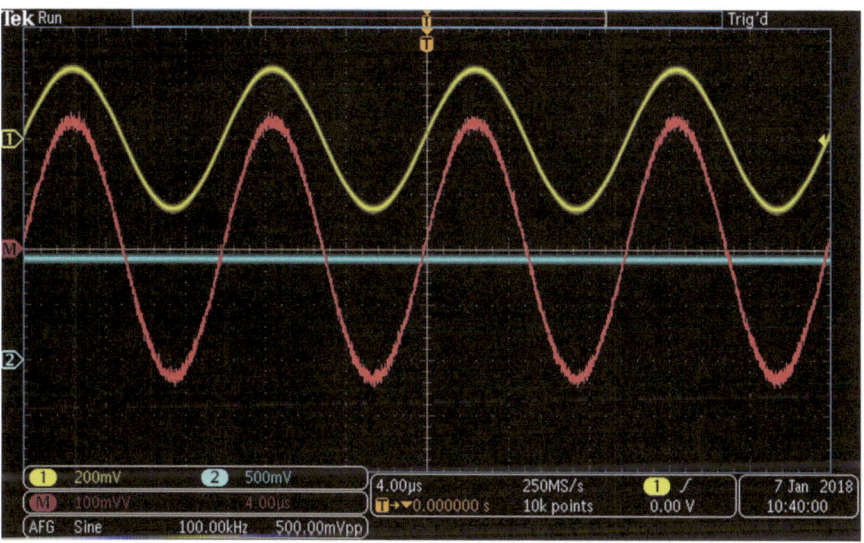

Fig. 4.44 Multiply. (Author's screenshot)

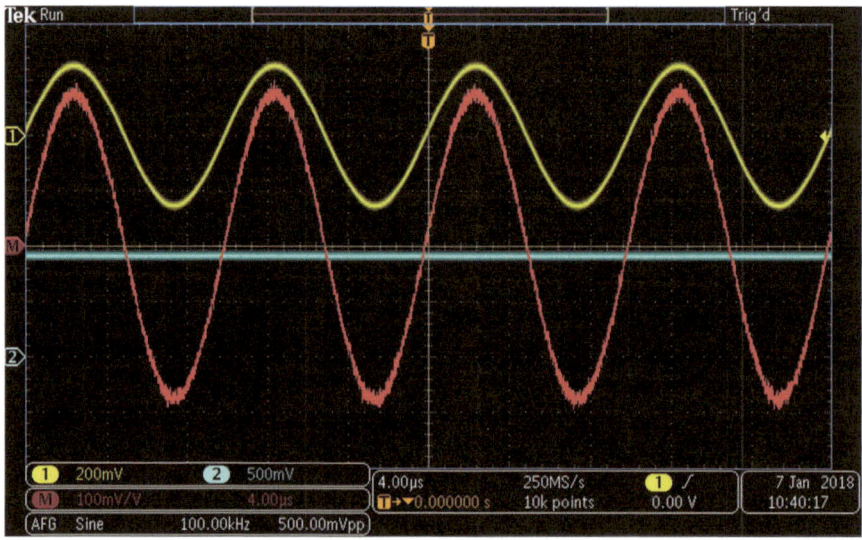

Fig. 4.45 Dividing by one-amplitude value

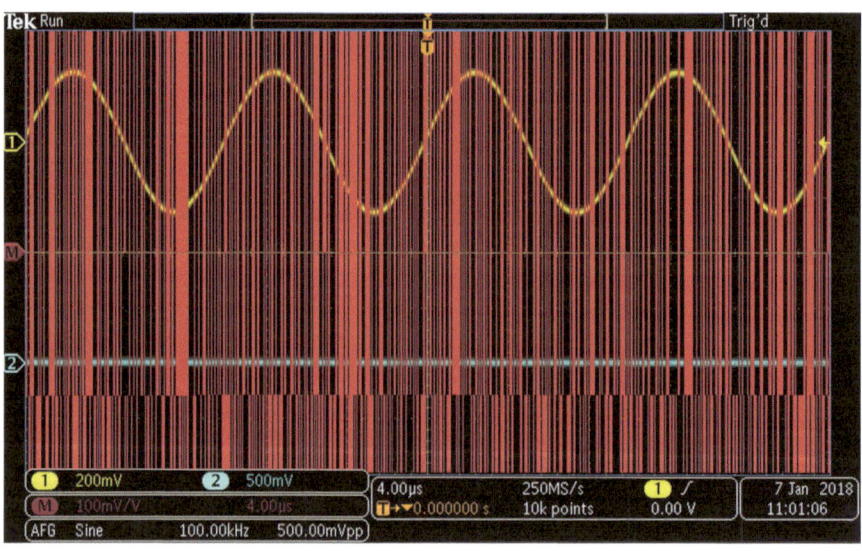

Fig. 4.46 Dividing by a zero-amplitude value. (Author's screenshot)

In the time domain, the horizontal X-axis is denominated in units of time. A periodic signal, triggered so that successive waveforms coincide to form a single, coherent image, appears in the display. The Y-axis is denominated in units of amplitude, usually volts or fractions or multiples thereof.

In the frequency domain, the Y-axis also represents amplitude, but in contrast to the time domain, the scale is usually logarithmic rather than lineal. It is denominated in decibel units of power. The X-axis, rather than time, shows a user-defined range of frequencies, and consequently the display has a totally different appearance. In the vertical FFT menu on the right, the top menu selection permits the user to set the Source by turning Multipurpose Knob a. The choices are Channels One through Four and Reference signals R1 through R4.

For now, we'll connect a sine wave from the internal AFG to Channel One. Be sure that Channel One and AFG are both on. The time domain and frequency domain sine wave signals are shown in the same display. This is why the instrument is known as a Mixed Domain Oscilloscope (MDO). This feature, as we shall see, is very useful in troubleshooting as well as in power quality evaluation.

The sine wave in the frequency domain consists of a single large spike. It is the fundamental, with no harmonics because the waveform is sinusoidal. The spike is difficult to see because it exactly coincides with the grid line at the left edge of the display. What you see whenever the oscilloscope is operating in the frequency domain is a very prominent, rapidly fluctuating horizontal line close to the bottom of the display. At first you might think that these are harmonics. But no, harmonics are regularly spaced sinusoidal spikes that are multiples of the fundamental.

This chaotic portion of the display is the noise floor of the oscilloscope, always present in a frequency domain display regardless of the quality and condition of the instrument. It is a consequence of the fact that due to thermal activity in any conductive body that is not at absolute zero temperature, random motion of the charge carriers translates to a very small finite voltage that appears across leads that are affixed to the body any finite distance apart. This is particularly true of resistive components including semiconductors within any measuring instrument.

The fact that the fundamental coincides with the edge of the display can be resolved by pressing the Freq/Span button that is under the RF button. Don't press RF. Doing so will disable all analog channels along with the AFG, which is connected to Channel One. RF can be used for a frequency domain display if the BNC cable that carries the AFG signal is swung over to the RF input, using an RF adapter, but the analog sine wave is not displayed. Accordingly, we have to say that the mixed domain feature is not available when RF is active.

Returning to Analog Channel One with the sine wave at the input and the oscilloscope in the Math>FFT mode, with Freq/Span pressed so that the vertical menu is present at the right side of the display, we see that the Center Frequency is 1.50 GHz. To bring the fundamental to the center of the screen, we need to set this figure so that it is the same as the AFG sine wave frequency, which is currently at the default 100.00 kHz value, as shown in the AFG bar at the bottom of the display. (Or, you could go back to AFG > Waveform Settings to discover and/or change the frequency.) Returning to AFG > Freq/Span, use Multipurpose Knob a or, better, the number pad to set the Center Frequency to the AFG's 100.00 kHz. In Math>FFT, to bring the fundamental to the center of the screen, horizontal should be 1.00 MH and 12.5 MHz/div. That way, the center frequency is slightly offset from the Y-axis so that it can be clearly seen.

Less than advanced users at times have difficulties getting a waveform to display in a meaningful way. The best approach is to make sure that Center Frequency, Span and Start, and Stop frequencies are appropriate for the signal that is to be displayed. These values are interrelated and depending on the type of instrument, when one of them is adjusted others may automatically fall in line. A square wave in time and frequency domains is shown in Figs. 4.47 and 4.48.

So far we have discussed the procedure for displaying signals in the frequency domain. How it relates to the time domain display of the same waveform is a large, complex subject about which thick volumes have been written.

While a comprehensive mathematical understanding of the back-and-forth interplay between time and frequency domains is not essential to operate the FFT and RF features in a DSO, most users eventually conclude that a theoretical background has value on its own merits.

More on Fourier

Joseph Fourier (1768–1830) vastly expanded our understanding of possible ways to represent, measure, and interpret waveforms. Previous thinkers from Pythagoras to Newton and many in the centuries between had observed oscillating energy in ocean

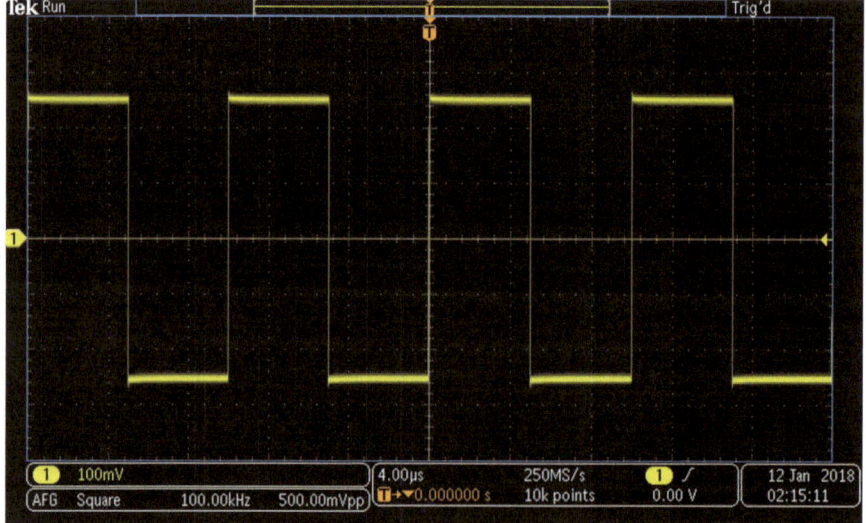

Fig. 4.47 Square wave in time domain. Harmonics are not visible in this time-domain representation of a square wave. (Author's screenshot)

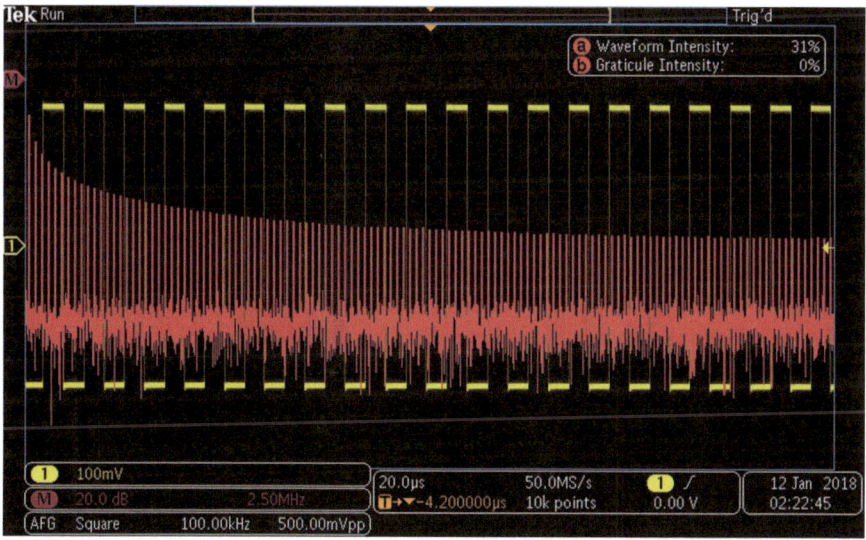

Fig. 4.48 In frequency domain mode, the oscilloscope displays the many harmonics present in a square wave. They are due to the extremely fast rise and fall times, which have high-frequency properties. (Author's screenshot)

waves, music, and economic cycles, but it was Fourier who quantified waveform behavior in ways that have resonated for the past three centuries and are applicable in today's electronic environment.

Fourier's paper, *The Analytical Theory of Heat* (1822), available in English translation at Google Books, states that "any function of a variable, whether continuous or discontinuous, can be expanded in a series of sines of multiples of the variable."

What this means is that a signal (other than sine) is composed of more than one sine wave. According to Fourier theory, this non-sinusoidal signal can be freely translated back and forth with no loss of information between the time domain and the frequency domain.

A complex non-sinusoidal signal displayed in the time domain typically has an irregular appearance, and, unless it is a very near match to the sine wave, it is easy to see there is electrical or other energy that is separate from the fundamental. But the frequencies at which this energy occurs are not evident in the time domain display.

Pressing FFT, the oscilloscope enters its frequency-domain mode, and the user can see the spectral distribution of all energy contained in the signal. Typically (unless it has been filtered out) there is a large spike at the center frequency. This is known as the fundamental, and there is a succession of smaller spikes diminishing in amplitude with spectral distance from it. This display clarifies the concept of bandwidth in signal processing, which, as the name implies, is a measure of the difference between upper frequency and lower frequency of a continuous output. It is measured in hertz and multiples thereof. To obtain a meaningful number, it is necessary to specify an amplitude at which the signal is considered present. Otherwise, since the amplitude tapers off and approaches zero at the frequency limits, the bandwidth could be considered infinite. The amplitude limit is variously considered 3 dB or 10 dB at each end, depending upon the application.

Fourier did not envision his theory as applicable to oscillating electrical (much less electromagnetic) waves. He was concerned with heat transfer about an iron ring, within and dissipating from a sphere or traveling through an infinite prism. Currently the Fourier Transform, which can denote either the end result or the mathematical operation, has been found to apply in a great many areas, including electronics, acoustics, mechanical engineering, statistics, and others. In all cases, the oscilloscope, equipped with appropriate transducers, is the principle instrument for performing the Fourier Transform and displaying the results.

The mathematics involved in realizing the translation of time domain to frequency domain (Fourier analysis) and back (Fourier synthesis) is very complex and not feasible for most purposes. Fortunately, based on the work by Carl Friedrich Gauss (1777–1855), James Cooley and John Tukey developed the very useful Fast Fourier Transform finalized in 1965, greatly simplifying the Fourier Transform as a working procedure. FFT is incorporated in contemporary DSO's, and it is capable of Fourier analysis of any periodic signal at the instrument's input. The user has only to press the FFT button and a frequency domain display of the original time-domain signal appears on the screen instantly.

FFT uses an algorithm to factorize the Discrete Fourier Transform (DFT) matrix into a product of mostly zero factors, and this accounts for an increase in speed of orders of magnitude, the exact amount depending upon the complexity of the original non-sinusoidal signal.

FFT was never patented because unlike Cooley, Tukey did not actually work for IBM. Immediately, the innovation went into the public domain, then becoming available for general use.

A consequence of the widespread use of FFT is that spectral analysis has become available for electronics engineers, technicians, and students. It is used on a daily basis in research, product development, and repair. A graduate degree in mathematics is not required to actually perform Fourier analysis and synthesis and to look deeper into waveform behavior, which in the post-quantum age is widely thought to underlie all matter and energy in our continuously expanding universe.

Chapter 5
Sampling, Memory Depth, and Bandwidth

Abstract Oscilloscope memory depth and sampling rate problems. Sampling rate, memory depth, and bandwidth are interrelated. It is necessary to understand how they work together if the oscilloscope is to be used efficiently for debugging new projects. For example, a harmful operating mode may be the result of a runt waveform or other glitch caused by a power supply anomaly. With more memory, it is possible to maintain a high sampling rate over a longer period of time. The higher sampling rate translates to a better chance of finding a bad waveform. The oscilloscope's effective bandwidth is greater. But there is a downside to a large memory depth. Under certain conditions it slows the oscilloscope. If the central processing unit cannot keep up with the demands of a deep memory, there will be more dead time.

Practical methods for evaluating and working around this parameter are suggested.

To become an adept oscilloscope user, one's knowledge base must include these interrelated acquisition concepts:

- CPU-Based Architecture
- Bandwidth
- Acquisition Memory
- Display Memory
- Sample Interval/Sample Rate
- Time/Division Settings
- Update Rate
- Memory Depth/Record Length
- Segmented Memory
- Resolution
- Triggered Sweep

The list is not inclusive. We could add extensively to it without exhausting the subject. But this should be a good start. We'll define each concept, show how they relate to one another, and state how they are applicable to processing, measuring,

© Springer Nature Switzerland AG 2020

D. Herres, *Oscilloscopes: A Manual for Students, Engineers, and Scientists*,

https://doi.org/10.1007/978-3-030-53885-9_5

and display of analog signals. We'll discuss digital processing, at once simpler and more complex, in Chap. 6 (Horowitz and Hill 2015).

CPU-Based Architecture

The contemporary digital storage oscilloscope (DSO) is built around a central processing unit (CPU) much like a personal computer. Within a DSO, the CPU occupies a strategic position, between analog channel inputs and flat-screen display. Digitized signals pass through the CPU, which performs logical, mathematical, control, and input/output operations in accordance with instructions hardwired onto the silicon substrate of an integrated circuit.

In a DSO as in a computer, the IC may also contain memory and peripheral interfaces. In a small package, the CPU is very fast and efficient, performing a multitude of functions as required by the modern digital oscilloscope, particularly in view of the high-frequency signals that are processed, and the great bandwidth demanded by many applications (Hickman 2001).

All this being said, we have to acknowledge the fact that the CPU can present a bottleneck or constraint that in certain situations actually slows down the oscilloscope to the extent that it may skip over needed data.

We put DSO's to various uses depending upon our specific priorities. Foremost among them is what is known as debugging. This colorful word has an interesting history. It was first used by Admiral Grace Hopper in the World War II era. Working on a Mark II computer at Harvard University, her coworkers discovered that the cause of a computer malfunction was a moth stuck in one of the thousands of relays. She commented that her colleagues were "debugging" the system.

That is what engineers do to electronic prototypes and computer networks that are not performing as expected. Often a defect is far less visible than a moth in a relay. For this reason, troubleshooting may involve highly abstract reasoning and the use of sophisticated test and measurement equipment. The tool of choice is of course the oscilloscope.

The procedure is to monitor the electrical output at appropriate points in hope of observing a glitch, runt, or other anomaly, often transitory and low amplitude, hidden in the rush of pulses and bursts in the ongoing data stream. It is in the interest of capturing these transients that Keysight Technologies (formerly Agilent and before that HP) has gone in a different direction, essentially doing an end-run around CPU-based oscilloscope architecture without of course discarding the actual CPU.

The debugging process begins by attempting to capture an anomalous infrequent event. The odds of doing this vary inversely with the amount of dead time. There is a shorter amount of dead time when the update rate is faster, and this is achieved by Keysight in deploying what they call Megazoom, which is in contrast to conventional CPU-centered oscilloscope architecture. Tektronix engineers for their part have developed Wave Inspector, and we'll be looking at that later. For now, some basic concepts:

Bandwidth

Many electricians (who are not electronics technicians) are puzzled by the notion of bandwidth. It implies that a signal can have more than one frequency. How is this possible? The truth is that some signals do consist of more than one frequency, whereas others, such as a pure sine wave, do not (Horowitz and Hill 2015).

In actuality, bandwidth can have a multiplicity of meanings, depending on the context. Most fundamentally, bandwidth is a measure of the difference between highest and lowest frequencies within a given continuous spectrum, generally in reference to a single unified signal. A complex, non-sinusoidal signal in the time domain decomposes into a finite or infinite number of sine waves in the frequency domain, each occurring at a specific, discrete frequency. This fact accounts for the phenomenon of bandwidth, which can be seen quite clearly when, with a non-sinusoidal signal at an oscilloscope analog channel input, the user presses Math>FFT. A mixed-domain instrument displays time domain and frequency domain graphs simultaneously. In the frequency-domain display, the precise spectral distribution of the signal is shown, and amplitude and time values appear when the cursor button is pressed and held to bring up the cursor menu.

The spectral components taper off in amplitude until they eventually disappear below the noise floor of the instrument. To obtain a meaningful bandwidth metric, an amplitude (power as represented logarithmically on the Y-axis) has to be specified, and by convention this is generally 3 dBm or 10 dBm depending on the application.

Another use of the word Bandwidth is as a figure of merit for various types of electronic equipment including the oscilloscope. What it refers to is the highest-frequency signal that can be displayed by the instrument without excessive attenuation and loss of triggering.

Attenuation at high frequencies is caused among other thing by parallel capacitive reactance. Capacitive reactance varies with frequency as well as capacitance of the circuit including all cabling, components, circuit boards, etc. Even unintended capacitance can severely limit high-frequency performance. Series inductance also causes attenuation in oscilloscope input circuitry. It is increasingly difficult to eliminate attenuation at high frequencies, and that is why high-bandwidth instrumentation is very expensive.

One method for reducing noise in a signal is to intentionally temporarily reduce the bandwidth of the instrument. Since noise is a very broadband phenomenon, it is significantly limited because the portion of it above the reduced cutoff frequency is eliminated. To reduce bandwidth, press the button for the channel input and press the soft key associated with bandwidth. A reduced level can be chosen.

Still another use of the word Bandwidth is as a measure of network speed or maximum data throughput in bits per second. One method for measuring network speed is to transfer a large file from one system to another system and measure the elapsed time. Throughput is quantified by dividing file size by transit time. However, since this does not accurately take into account large variations in latency, window size, and system bandwidth, it cannot always be considered definitive.

Acquisition Memory

In a DSO, acquisition memory communicates continuously and at a speed that is high enough to keep up with the waveforms flowing in from the analog inputs. And through the ADC, acquisition memory looks back at the oscilloscope front end, which amplifies or attenuates, conditions, and digitizes waveform data prior to processing and display.

To demonstrate, with a signal applied to a channel input, press Acquire to see the menu choices. First in the menu is Mode and the default is Sample, the simplest of the acquisition modes. It retains the first sampled point from each acquisition. Because it is the default mode, it is by far the most commonly used.

Peak Detect Mode works well for high-frequency glitches, but it can be used only with real-time, non-interpolated sampling. It uses the highest and lowest of all the samples contained in two consecutive acquisition intervals.

High Resolution Mode also works only with real-time, non-interpolated sampling. It calculates the average of all samples for each acquisition interval. High Resolution Mode displays a higher-resolution, lower-bandwidth waveform.

Envelope Mode incorporates Peak Detect for each individual acquisition. It captures the highest and lowest record points over all acquisitions.

Average Mode is highly effective for reducing noise in a signal. It uses Sample Mode for each individual acquisition. It calculates the average value for each record point over a user-specified number of acquisitions.

To get a good sense of how these modes work, go back to AFG and then scroll down to noise. With the noise signal displayed, press Acquire once more and successively activate the Acquisition Modes. Of particular interest is Average. As increasing numbers of acquisitions are averaged, the amplitude of the noise signal diminishes, until at 512 all you see is a zero-amplitude signal in the Cartesian coordinates. A similar experiment can be performed by scrolling in AFG to a signal such as sine, and adding various percentages of noise.

The next menu selection in the Acquire menu is Record Length. Using Multipurpose Knob a, this parameter can be varied from 1000 points to 10 million points. Reasons for setting Record Length at less than maximum will be discussed in Chap. 6 when we consider this strategy in the context of digital signal analysis.

After Record Length is FastAcq, which can be toggled Off (default) or On. When it is turned On, the Waveform Palette submenu appears. Choices are Temperature, Spectral, Normal, and Inverted.

One of the purposes of FastAq is to provide high-speed waveform capture and display. It is very useful in finding elusive signal anomalies. FastAcq reduces the dead time between waveform acquisitions so that glitches, runt pulses, and other transient events are captured and displayed. Another useful FastAcq capability is that it can display waveform phenomena at an intensity and coloration that indicates their relative rates of occurrence. Temperature, Spectral, Normal, and Inverted are user-selected preferences that govern how colors are used to identify events that occur more often. Infrequent anomalies can also be identified. All of this is enormously useful, as we shall see in Chap. 6, in developing prototypes and pinpointing failed digital circuits in the repair shop.

Delay, the acquisition menu selection that follows Fast Acq, is activated by pressing the associated soft key to toggle this feature on. It is used in conjunction with the next menu selection, Horizontal Position. With Delay turned on, you can turn Multipurpose Knob a counterclockwise to increase the delay. This has the effect of moving the trigger point to the left and beyond the acquired waveform. Then adjust the horizontal scale knob so as to acquire greater detail around the area of interest.

With Delay on, the trigger point moves away from the horizontal expansion point, which remains at the Y-axis. When the trigger point shifts off screen, the trigger marker points toward the trigger point. The purpose of Delay is to acquire waveform detail as it is temporally separated from the trigger point.

Display Memory

Figure 5.1 shows the relationship between acquisition memory and display memory in a DSO.

The DSO is a great breakthrough compared to the purely analog oscilloscope that preceded it. The oscilloscope giants are Tektronix, Teledyne LeCroy, and Keysight Technologies, formerly Agilent. These major players in electronics instrumentation carefully watch one another, each of them looking to seize the high ground. Tektronix, following World War II, built on British radar circuitry to develop triggered sweep, which made the new Model 511 oscilloscope, shown in Fig. 5.2, viable in the new TV servicing environment.

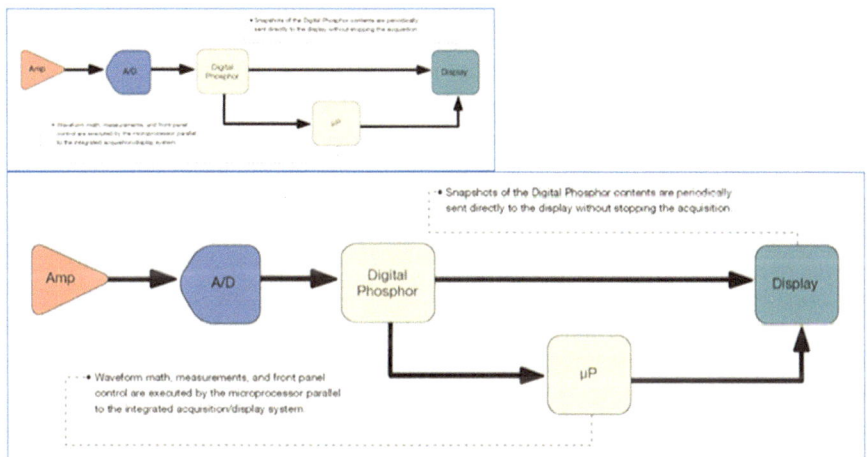

Fig. 5.1 A DSO block diagram. Acquisition memory is before the CPU and display memory follows it. (Tektronix)

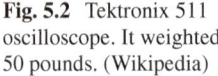

Fig. 5.2 Tektronix 511
oscilloscope. It weighted
50 pounds. (Wikipedia)

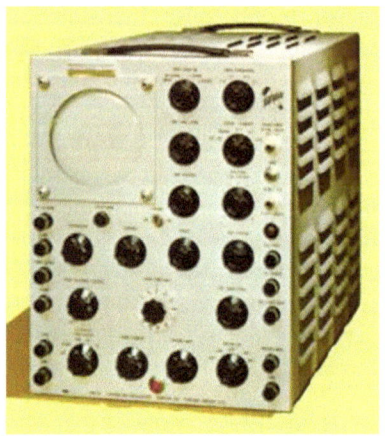

Soon LeCroy rewrote the book on oscilloscope engineering by replacing the old
analog design with today's technology – acquisition memory, digital CPU, and display memory connected to the now ubiquitous liquid crystal display. Walter
LeCroy's background was in high-energy physics research. From his position at
CERN, which operates the world's largest particle colliders, he brought his highly
developed expertise in high-speed digitation back to the United States, where with
colleagues he founded LeCroy Corporation, now Teledyne LeCroy. That all made
for a more complex but far more capable, efficient, and less costly instrument in a
relatively compact package. The user interface was intuitive and permitted much
tighter control and greater capability in waveform recall, measurement, and display.

Sample Interval and Sample Rate

A discussion of sample interval in a DSO has to begin with a definition of time duration of a waveform. This is user-determined, not a fixed amount for the oscilloscope
make and model. In the time domain, it is measured along the X-axis in seconds or
fractions/multiples of a second. By convention, in most instruments there are ten of
these horizontal divisions, and this space is always considered equal to the time
duration of the waveform. That being said, how is it possible that the time duration
of the waveform can be determined by the user? Quite simply, the user can adjust
the horizontal scale by turning the scale knob in the Horizontal section.

If you apply a sine wave from the AFG to an analog channel, it will be displayed
at the default frequency, typically 100.00 kHz. Then, turning the horizontal scale
knob counterclockwise, the sine wave oscillations will squeeze together and one
might conclude that the frequency has increased. This is not the case, as can be seen
in the AFG bar, where the frequency has remained at 100.00 kHz. Nor has the number of divisions changed. It remains at ten, five on either side of the Y-axis.

Similarly, the horizontal scale knob can be turned clockwise, and the waveform as displayed will be lengthened horizontally. Here again the frequency and number of divisions are unchanged. The Wave Inspector view shown in split screen format at the top shows clearly what is going on. The horizontal scale can just as well be varied by turning the small inner zoom knob in the Wave Inspector section. In all cases, the total time duration is ten times the horizontal time scale.

Another fixed parameter of a DOS is the sample rate, usually printed on the front panel near the instrument's overall bandwidth. If a given oscilloscope has a 5GS/s sample rate (200 per sample interval) and the horizontal scale is 20 μs/div, then the memory depth required will be one million points.

Waveform Update Rate

Digital circuits are prone to what is known as metastable states, which are output conditions where glitches or setup and/or hold-time violations are present. The oscilloscope user is often tasked with locating and displaying these anomalies so that they can be correlated with physical dysfunctions in prototypes as well as equipment brought into the shop for repairs or larger systems including networks. That would seem straightforward, but sometimes it is a time-consuming search for a needle in a haystack. The needle is small, the haystack is orders of magnitude larger, and it is an unavoidable fact of life that all digital oscilloscopes must contend with a blind spot following each acquisition as it is processed. Figure 5.3 maps out this problematic reality.

This dead time is an inherent characteristic of all digital oscilloscopes. During dead time, the signal(s) at the instrument input(s) are not displayed. Accordingly, random and infrequent events may be missed. The good news, however, is that the more often an oscilloscope updates waveforms, the higher the probability of observing harmful anomalies.

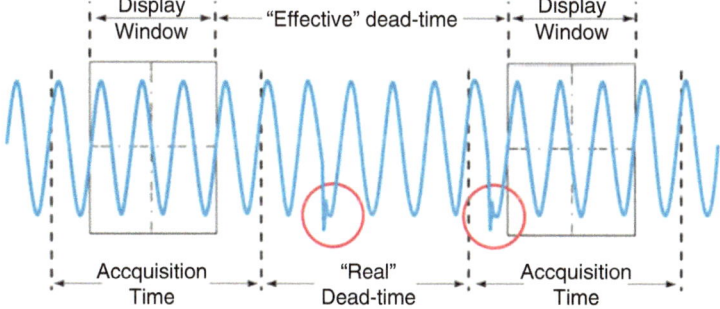

Fig. 5.3 Acquisition time starts before and ends after the display window as defined by the oscilloscope's display memory, so effective dead time is even greater than you might think. (Courtesy of Keysight Technologies)

Update rate is measured in waveforms per second. Obviously, the faster, the better. Even a few thousand updates per second, however, may not be fast enough to catch an elusive transient if, for example, it happens only once in a million appearances of the signal under examination.

Another aspect in this discussion is the fact that in digital acquisitions, the update rate is actually slower than the banner analog rate. The procedure for quantifying dead time is to apply this equation:

$$\text{Percent of dead time} = 100 \times (1 - \text{UW})$$

Where:

$$U = \text{Oscilloscope's measured update rate}$$

and

$$W = \text{Display acquisition} = \text{Time base setting} \times 10$$

As you can see, the oscilloscope's time base setting is critical when it comes to figuring update rate, because it determines the acquisition display time window. If the user adjusts the time base to longer time/div settings, the instrument has to digitize longer waveforms.

If you need to know the oscilloscope's waveform update rate, you can check the instrument's specification. An alternative is to determine it empirically. This is done by observing the frequency of an output trigger signal, using an external counter.

Memory Depth/Record Length

An oscilloscope's memory depth is equivalent to the maximum possible record length (number of samples) for one acquisition. Are record length and memory depth the same? No! The way they are related is that the *maximum* record length is dependent upon the oscilloscope's amount of memory.

Some digital oscilloscopes permit the user to directly adjust the record length. In others, the record length is a value dependent on the sampling rate and the time per division, which is determined by the horizontal scale knob.

In an instrument in which the record length can be directly adjusted, press the Acquire button, which brings up the horizontal Acquisition menu below the display. With a signal, such as a waveform from the internal AFG present at an active analog input, press the soft key associated with record length. The vertical record length menu appears at the left of the display and the user can, by turning Multipurpose Knob a, select the desired record length quantified in points. The choices are 1000, 10 k, 100 k, 1 m, 5 m, and 10 m. The chosen record length is displayed in the acquisition information bar and remains there, as well as in the Acquisition menu even after you cancel out of the record length menu.

What concrete effect does altering the record length have on the way a signal is displayed? Selecting the correct record length for a given application optimizes the level of detail displayed. A record length of 2000 points is adequate to display a stable sinusoidal signal, but to uncover timing anomalies in a complex digital data stream, the record length should be increased to 1 m points or more.

Wave Inspector

Wave Inspector greatly enhances oscilloscope functionality when it comes to searching for waveform anomalies that may be causing poor performance in electronic equipment. It is particularly useful for finding transients, runts, and all sorts of harmful electrical energy in digital signals. We'll have much more to say on this topic in Chap. 6 when we discuss digital signals in detail. For now, we'll see how Wave Inspector can bring out detail in analog waveforms, and we'll look at some techniques for getting the most out of this valuable tool.

To see how Wave Inspector works on analog signals, from the internal AFG or other appropriate source, apply a sine wave to the Channel One analog input. The AFG information bar shows the default parameters. The signal's frequency is 100.00 kHz, and the amplitude is 500.00 mV_{pp}, in other words half a volt. These values can be changed by pressing the soft key associated with Waveform Settings, but for now they are fine.

To activate Wave Inspector, press the button identified by the magnifying glass icon, as shown in Fig. 5.4.

Figure 5.5 shows the Wave Inspector screen with a sine wave displayed.

The split-screen format makes Wave Inspector highly user-friendly. In the upper display, we see the sine wave as applied to the analog input. Notice the pair of brackets. The portion of the waveform that is enclosed between the brackets is shown in the lower display. The user can move the brackets by turning either of the concentric knobs in the center of the Wave Inspector section. The smaller inner knob moves the brackets closer together or farther apart to control the degree of zoom and hence the amount of detail shown in the lower display. The larger outer knob moves the pair of brackets from side to side, panning throughout the overall waveform so that varying portions of the time base are displayed. You can see how useful this tool can be in bringing out transient anomalies that would otherwise be difficult to discern.

Panning can also be automatic, permitting the user to focus on detail without twirling a knob as the overall waveform is scanned. To illustrate, use the inner zoom knob to bring the brackets close to one another, almost touching, and then press the Wave Inspector button that is identified by the movie start-stop icon. The brackets move slowly across the time base, allowing the waveform to be viewed in very great detail. This scanning can be stopped when something of interest is found. If the brackets are moved farther apart, the motion accelerates, permitting this capability to be customized as desired.

Fig. 5.4 Wave Inspector is a tool that permits the oscilloscope user to view brief, infrequent transients that could otherwise be difficult to find in a long waveform record. (Tektronix)

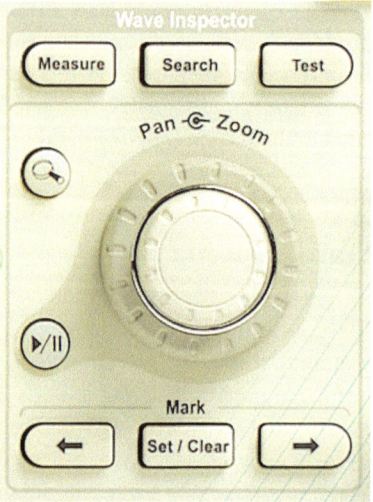

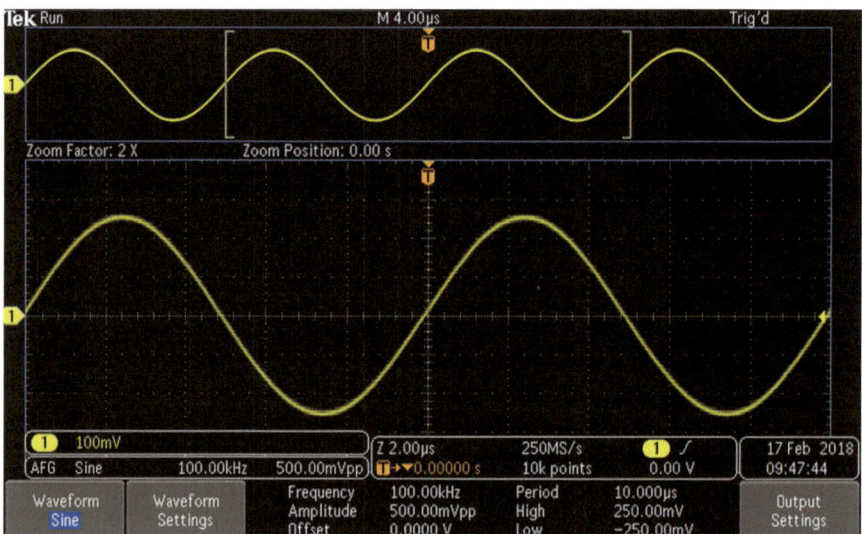

Fig. 5.5 Wave Inspector looks at a sine wave. (Author's screenshot)

Along the bottom of the Wave Inspector section on the front panel are the three Mark buttons. The center button is labeled Set/Clear and on either side of it are Back and Forward buttons. These controls allow the user to place marks at will within a waveform record.

There are a number of reasons why you may want to insert a marker. If you find a suspected anomaly or something just doesn't look right and warrants further investigation, you can place a marker at the location, and there will be no need to do the search all over. A good plan is to scan the waveform and place markers to get an overview and then later perform a final analysis. Another motive in placing a marker might be to use it to pinpoint a location within a saved screenshot that is to be emailed to a colleague or to tech help.

When you see an event that you want to mark, pan the brackets so that it is centered between them. Then press Set. Markers appear as white triangles in both upper and lower displays. Pressing Clear removes the markers in both displays. It is a simple matter to pan to other locations and set additional markers.

If you power down the instrument and start it up the next day, the markers will be right where you left them. However, if you do a default setup, they will be gone.

Notice as you pan the Wave Inspector brackets, the markers in the upper display remains motionless while the markers in the lower display move with the brackets.

So far we've discussed the process of manually placing markers. The user simply decides where they should be located and places them using the Set button. Additionally, you can specify types of events and Wave Inspector will search the entire acquisition and automatically mark each occurrence of the event.

For example, if a waveform consists of a large number of pulses, the user can direct Wave Inspector to mark each one. Then, without traversing large amounts of dead time between them you can press the Previous and Next buttons to place the markers and corresponding time bases within the Wave Inspector brackets. To perform this task, press the Search button above the concentric knobs. This brings up the horizontal Wave Inspector Search menu at the bottom of the display. When you press the soft key that is associated with the first menu selection, the vertical Search menu appears and the first menu selection allows you to toggle Search On and Off. When Search is Off (default), the user can clear all markers, copy search settings to trigger, copy trigger settings to search, or convert automatic markers to user markers. (User and markers, as we have seen, are solid white triangles. Automatic markers are white with dark interiors.)

Toggling Search On, the horizontal menu selections are Search, Type, Source, Slope, and Threshold. The default Search and Threshold Type is Edge. Pressing the soft key, the vertical Search Type menu appears at the left of the display. Search Types are:

- Edge
- Pulse Width
- Time Out
- Runt
- Logic
- Setup and Hold
- Rise/Fall Time
- Bus

  We'll have a lot more to say about these search types, especially Bus, in Chap. 6, Digital and Logic Circuits.

  Source permits the user, by means of Multipurpose Knob a, to select a source, analog channel, memory, reference or digital channel.

Slope is rising, falling, or both. Threshold Voltage, which determines where the automatic markers are placed, can be selected by turning Multipurpose Knob a. Once the automatic markers have been placed, you can press the Previous and Next buttons to navigate among markers.

Segmented Memory

In an oscilloscope, segmented memory can optimize acquisitions, making possible the capture of elusive signal details. In viewing low duty-cycle pulses, segmented memory captures only a small selected time segment adjacent to each pulse, while disabling the oscilloscope's digitizers during dead time. Viewing low duty-cycle events like laser and radar signals, sonar bursts, or packetized digital data can be problematic for oscilloscope acquisition because the signals are very brief with long time-base intervals between them. The memory needed far exceeds available resources. A solution is segmented memory, which enables the oscilloscope to capture huge numbers of successive single-shot waveforms while ignoring the dead time between them.

The user can manually set the time base to a slow time/div rate in an attempt to capture long time spans and numerous digital packets, but soon the oscilloscope's maximum time span and sample rate are exceeded. We lose horizontal and vertical waveform detail.

Segmented memory selectively captures the signals of interest without expending valuable memory on irrelevant dead time, during which the digitizer is intentionally disabled.

Segmented memory is essential in high-energy physics projects involving very demanding laser pulses, which are vanishingly brief and separated by long intervals of electrical silence.

Additionally, viewing radar and sonar bursts as well as serial bus data, as we shall see in Chap. 6, are by their temporal nature demanding applications that often require segmented memory to isolate suspected transients and anomalies.

Resolution

Most modern digital oscilloscopes have eight-bit resolution. This specification relates to the analog to digital converter (ADC), and it quantifies the instrument's ability to discriminate between two values that are in close proximity in time (horizontal resolution) and amplitude (vertical resolution). We are usually more interested in vertical resolution.

The first digital oscilloscopes, appearing in the 1980s, typically had 6-bit ADCs. The resolution of an ADC is 2^n where n is the number of bits. Thus, a 6-bit ADC was capable of discerning 64 vertical quantization levels. Later oscilloscopes had 8-bit ADC's, so they could display 256 quantization levels. That, today is the standard, although instruments with greater resolution are available.

It must be emphasized that resolution is not the same as accuracy. Accuracy denotes the degree of fidelity of a displayed waveform with respect to the signal under investigation; let us say at the output of a transistor on a circuit board. It could be compromised by poor probing, to mention one example. Vertical resolution is a measure of the ability in an instrument to differentiate between amplitude levels that are located close to one another with respect to the Y-axis. It would be quite possible in an oscilloscope to have high resolution but low accuracy. On the other hand, low resolution would always impose an upper limit on accuracy.

Resolution is very important when you are attempting to display small differences in large-amplitude signals, such as AC ripple in a utility voltage, or in the high-voltage DC bus in a 480-volt variable-frequency AC motor drive.

If you are having problems displaying small signal differences in an electrically noisy environment or when the overall electrical dynamics are challenging, there are helpful techniques that are available.

You probably won't want to open the oscilloscope enclosure and replace the ADC with a higher-resolution upgrade. (That would involve replacing some other components also.) But there are other less radical system-wide remedies that will make a big difference. We'll start with probing.

You can enhance small-signal resolution by reducing noise from outside as well as inside the oscilloscope. To evaluate the effectiveness of changes made in probing as well as elsewhere, a good plan is to take some before and after screenshots. Using the standard 10-1 probe that came with your oscilloscope, hook onto a non-fluctuating, low amplitude, fairly high-frequency signal. It is instructive to look at it in the time domain as well as in the frequency domain. As you know, noise in the frequency domain appears as an irregular rapidly fluctuating roughly horizontal line that is known as the noise floor. On a USB flash drive, save this screen for later reference.

Then, save the same signal as displayed in the time domain, also for later reference. Noise in the time domain appears as a thickening of the trace.

By way of comparison, access a sine wave from the internal AFG. Press Output Settings and you will see in the second menu selection that you can add various percentages of noise by turning Multipurpose Knob a. (Zero percent, of course, is the default.) When you add 10 percent noise, the trace thickens quite noticeably.

Going back to the signal under investigation and looking at it in the time and frequency domains, there is no way to ascertain the amount of noise that is an inherent part of the signal, that is coming in on the probing, and that is the instrument's inevitable noise floor. But using the techniques outlined below, we will soon discover the greatest noise source.

Reducing Noise

To perform high-resolution measurements of weak signals, we must endeavor to reduce external and internal noise while maximizing signal amplitude. The general-purpose 10-1 probe, while ideal for most oscilloscope tasks, is not suitable for picking up a very low-amplitude signal that may be compromised by external and/or internal noise. Attenuating voltage probes reduce the input signal by creating a voltage divider with the input impedance of the oscilloscope. The familiar 10-1 probe reduces the signal's voltage by a factor of 10. The oscilloscope in its front-end circuitry compensates by amplifying the signal, but unfortunately it also amplifies whatever inevitable noise is generated by the probe and oscilloscope. This situation is highly unfavorable in the context of a weak signal that you are endeavoring to measure.

One solution is to use a probe that has less attenuation. Some probes have a slide switch or screw adjustment that permits the user to select the amount of attenuation.

Another solution is to use a differential probe, as shown in Fig. 5.6.

Rather than a ground-return lead, the differential probe has two hot probes that can hook onto separate terminals or wires that are referenced to but float above ground potential.

The differential probe applies to the oscilloscope's channel input only the difference between the two circuit nodes, and any common voltage is not conveyed to the oscilloscope input. This is known as common mode rejection. Since noise consists

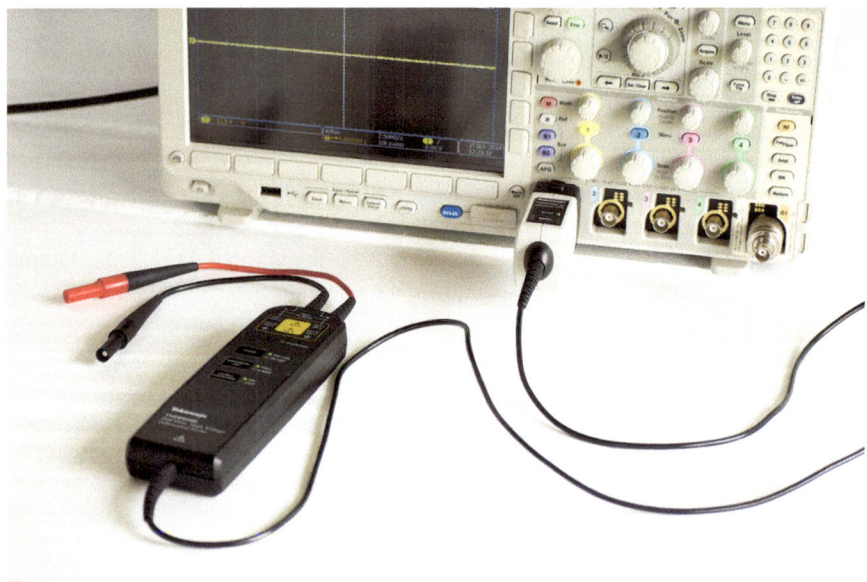

Fig. 5.6 Differential probe. (Judith Howcroft)

of more or less common electrical energy that is applied to both probes, it is for the most part eliminated. Accordingly, the signal-to-noise ratio is improved so that a small signal can be accurately displayed and measured in the oscilloscope.

A differential probe is fairly expensive ($850 or more), so it may not be available. Fortunately there are other remedies. A simple strategy is to isolate the oscilloscope and circuit under investigation from any outside sources of electrical noise. Short of building a Faraday cage in the shop or lab, you can locate the instrumentation away from offending items such as fluorescent ballasts, brush-type electric motors, switching power supplies, and the like.

Using a conventional (non-differential) probe, a good practice when acquiring small signals is to incorporate the shortest possible ground return lead consistent with the geometry of your test setup. That is because the ground return lead, through capacitive and inductive coupling, will pick up stray electromagnetic noise, adversely affecting the signal-to-noise ratio of the small signal that you want to display and measure. This capacitance and inductance may also resonate together to generate a spurious overlay. The length of the ground return lead should be minimized and connected to a ground potential as close as possible to the source of the signal of interest.

Another good strategy is to make use of bandwidth filtering, which may be built into active differential probe bodies. These probes sometime incorporate several bandwidth settings, permitting the user to eliminate noise, which is generally a wide-band phenomenon. Some active probes communicate to the oscilloscope that this filtering has been activated, whereupon the instrument switches on internal filtering, further improving the signal-to-noise ratio.

AC and DC Coupling

When you want to accurately measure a small AC signal that is riding on a large DC signal, as in checking a VFD DC bus, you may find the AC is difficult to quantify because it is scaled down in the presence of the large DC. (VFD inverters require high-quality, ripple-free DC to synthesize pulse width-modulated high voltage for the motor.)

Pressing the active channel button, the horizontal channel signal menu appears below the display. The first menu selection on the left is Coupling and the soft key toggles between AC and DC. DC coupling is the default and is used most of the time. It has no effect on the signal, so that AC and DC components are displayed just as they are applied to the oscilloscope input. DC should actually be termed AC + DC since DC coupling could be taken to mean that the AC component is suppressed.

AC coupling places a capacitance in series with the signal so that the DC component is blocked. This permits the user to see ripple in isolation, not scaled down in the presence of the higher amplitude DC. AC coupling persists when the instrument is power-cycled, so be sure to toggle it back to DC coupling when you are finished measuring ripple. Default Setup causes the channel to revert to DC coupling.

More on Noise

Going back to noise, which is often a bigger factor in small-signal resolution than scaling, the oscilloscope has two methods for noise abatement. One is bandwidth limiting and the other is signal averaging, also called waveform averaging. Both methods take advantage of specific noise characteristics in order to separate it out from the authentic signal. Bandwidth limiting takes advantage of the fact that noise is a wide-band phenomenon, covering more spectrum territory than the signal of interest. Signal averaging, in contrast, takes advantage of the fact that noise is a random amplitude phenomenon, whereas the signal of interest is usually periodic and does not change over a short time span.

Most oscilloscopes possess the capability to temporarily limit their bandwidth, typically to either of two levels significantly below full bandwidth as specified for the instrument.

To see bandwidth limiting in action, activate AFG, dial up Sine Wave, and press Output Settings. Add 30 percent noise. That should be enough to cause the waveform to lose triggering. Then, activate the relevant channel by pressing the channel button. This brings up the horizontal channel signal menu, below the display. Press bandwidth, bringing up the vertical bandwidth menu to the right of the display. Cutting the bandwidth to 250 MHz has very little effect, but dropping it further, to 20 MHz, cleans up the signal considerably, although some thickening of the trace indicates the presence of a small amount of noise.

Bandwidth limiting is fairly effective in reducing noise, but some noise is always present in the remaining bandwidth. Another disadvantage is that this method cannot be used for high-frequency signals where more bandwidth is required to display them.

These downsides are avoided in signal averaging. To see it in operation, press Default Setup>AFG > Sine. Again, add 30 percent noise so that triggering is lost. Then press Acquire>Mode > Average.

The user can select the number of acquisitions to be averaged, using Multipurpose Knob a. The range is 2 to 512. At 512 the very fine trace indicates a noise-free signal, but it seems to pulsate indicating a small amount of amplitude instability. Go back to AFG and lower Add Noise to zero percent. Now, when you range up to 512 acquisitions to be averaged, the waveform no longer pulsates. Notice also that the trace is very fine (noise free), even more so than the sine wave coming out of AFG without signal averaging applied.

An Interesting Experiment

If you can obtain a programmable electronic keyboard, you can connect it to your oscilloscope, as shown in Fig. 5.7, and see how sounds from various instruments appear in time and frequency domains.

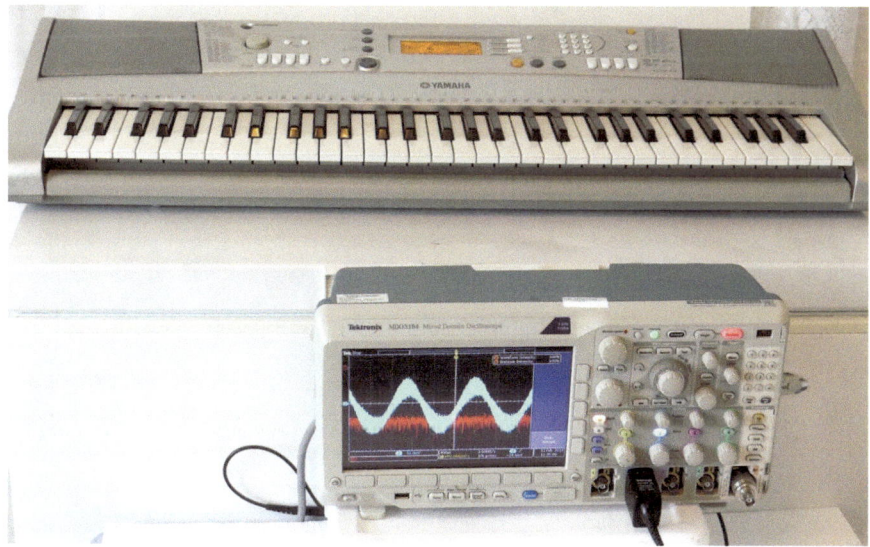

Fig. 5.7 Electronic keyboard. (Judith Howcroft)

A typical low-priced electronic keyboard is capable of synthesizing sounds synthesized by over 500 instruments including variations, and it is instructive to look at the waveform signatures. You can save them on a flash drive and assemble a fascinating library of waveforms in your computer. With practice hearing ambient sounds, you can anticipate their appearance in the frequency domain if not the time domain.

Most electronic keyboards have small internal speakers and an auxiliary mini-phone jack where headphones can be connected. Short of opening the enclosure and splicing in wires, you have to obtain a matching mini-phone plug and adapt its output so that an oscilloscope probe with hook tip and alligator clip attached to the ground return lead can be connected. Either that or cut off one end of a sufficiently long BNC cable and solder the wires to the mini-phone plug. Most of these plug-jack hookups have two hot wires and a common. You have to connect to one hot wire and the common.

These auxiliary plugs disable the internal speakers when they are connected, so you just have to live with the fact that you have to slide out the oscilloscope plug whenever you want to hear the note(s) being played. That is unless you want to go inside the keyboard and make changes.

You should fabricate a weighted object that will hold down a single key without disturbing adjacent ones in order to sustain the sound while you work the oscilloscope and keyboard controls.   Now we are ready to display some acoustic waveforms. The flute is a unique instrument because its upper register and lower register sounds differ not only in frequency but also in spectral distribution. A high-pitched flute note appears as a relatively pure sine wave, as shown in Fig. 5.8, while

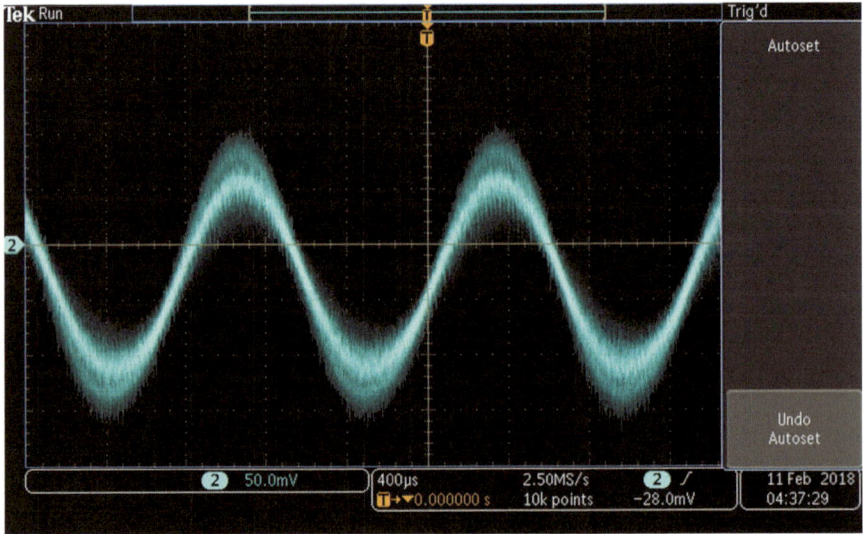

Fig. 5.8 High-register flute note in time domain. (Author's screenshot)

a low-pitched flute note displays in the time domain as a far more complex wave-form, as shown in Fig. 5.9.

In the frequency domain, as shown in Figs. 5.10 and 5.11, we see spectral distributions of these same flute notes.

The saxophone waveform, shown in the time domain in Fig. 5.12, is far more complex. Abundant harmonics, shown in Fig. 5.13, appear in the frequency domain, and they give the saxophone its unique characteristic sound.

Triggered Sweep

An oscilloscope's trigger function synchronizes signal and horizontal sweep. Prior to implementation of triggering, successive sweeps started at different temporal locations with respect to the signal. The result was multiple flickering images in the display and blurry thickened traces. The oscilloscope therefore required a high level of user expertise, and it was restricted to limited applications.

Following his military service in England during World War II, Howard Vollum returned to the United States and with colleagues founded Tektronix. Based on early radar technology, they introduced triggered sweep in their new line of oscilloscopes. One of these instruments, the Tektronix Model 511 oscilloscope, permitted a new generation of theoreticians to see electrical waveforms in real time and enabled a new generation of TV technicians to participate in the enormous post-war consumer electronics boom.

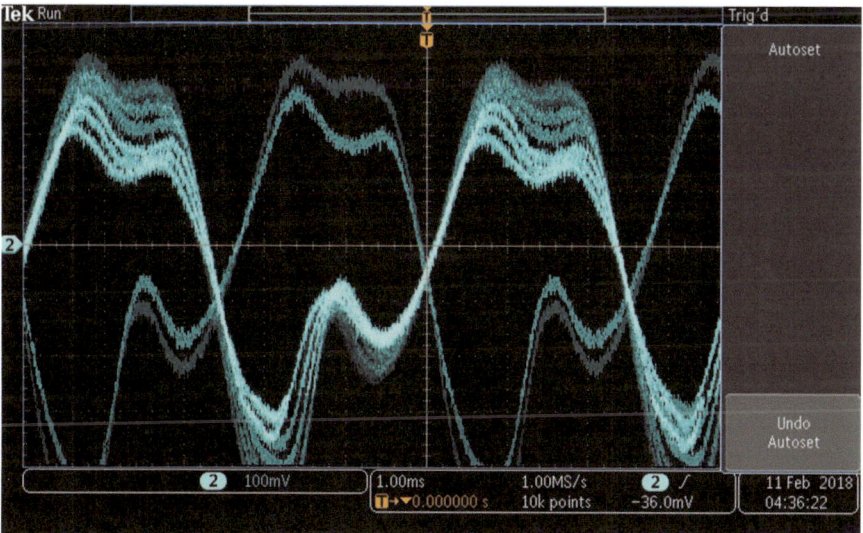

Fig. 5.9 Low-register flute note in time domain. (Author's screenshot)

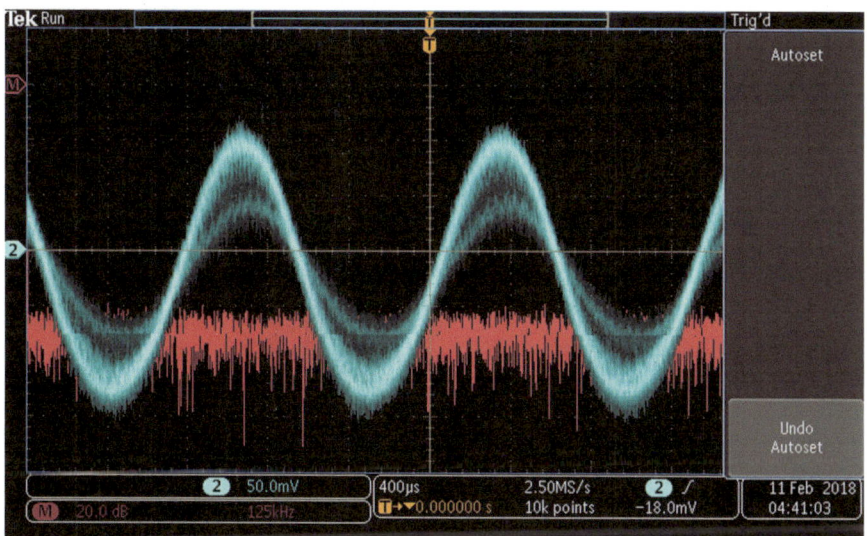

Fig. 5.10 High-register flute note in frequency domain. (Author's screenshot)

If an old-style analog oscilloscope with CRT could be built with an infinitely wide screen (assuming that the user could view it and the horizontal deflection voltage would be capable of driving the beam from side to side), there would be no need for triggering. A periodic, repetitive signal could display indefinitely. Real-world oscilloscopes have far smaller screens, and the trace quickly runs out of space.

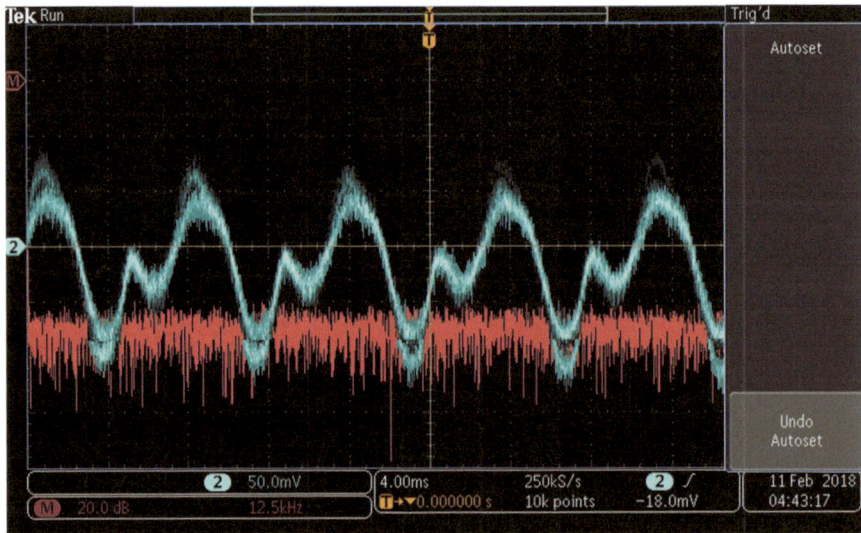

Fig. 5.11 Low-register flute note in frequency domain. (Author's screenshot)

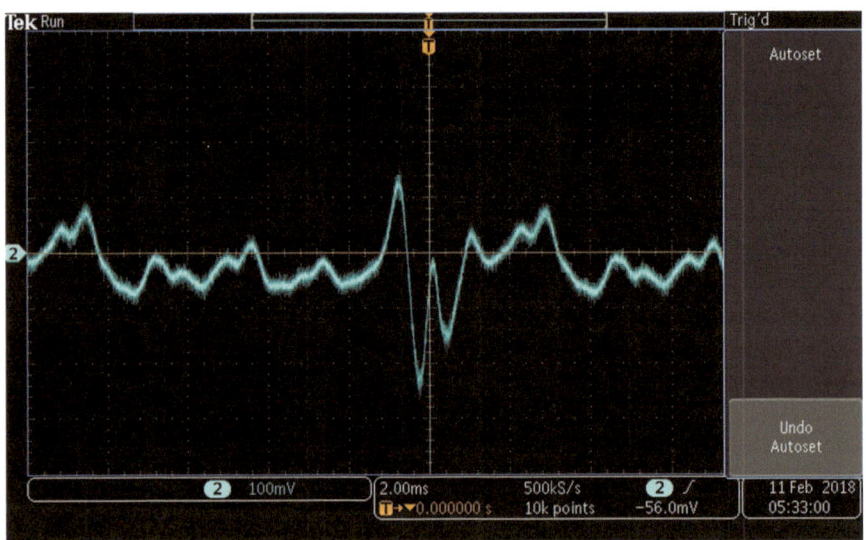

Fig. 5.12 Saxophone midrange note displayed in time domain. (Author's screenshot)

A CRT beam would instantly return to the top left of the screen, the beam blanked out as in TV during retrace, so as not to be a distraction. The problem in an oscilloscope attempting to display a repetitive signal would be that the waveform at the oscilloscope input and the trace as displayed would be unlikely to coincide, and this typically would happen a great many times per second, so there would be no continuity and the display would be an incoherent jumble.

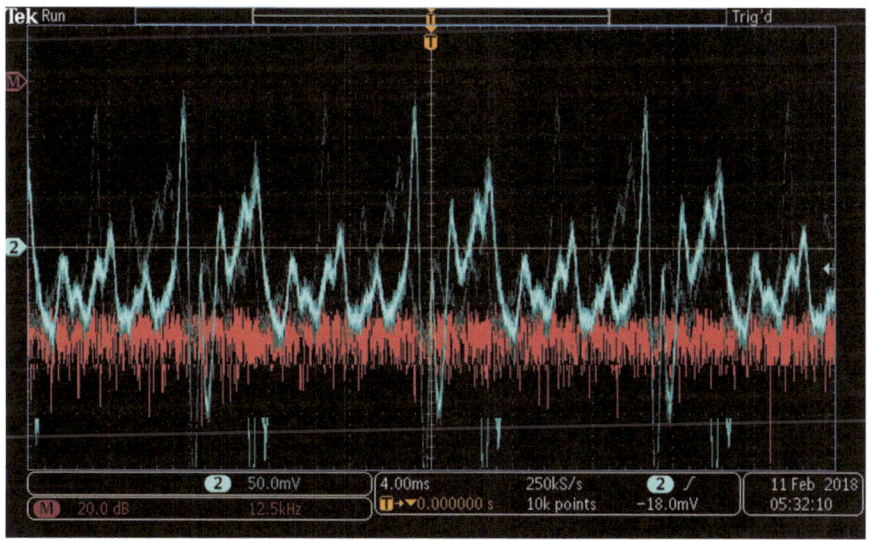

Fig. 5.13 Saxophone midrange note displayed in frequency domain. (Author's screenshot)

Tektronix figured out how to pause the acquisition for a very brief amount of time prior to a triggering event, the same for each waveform cycle, creating a stable display. Today this triggering event is user-defined, and it establishes a time-reference location in the waveform record. Meanwhile, the oscilloscope acquires additional samples, which are inserted in the pre-trigger segment of the waveform record.

The trigger point is indicated by an orange T icon at the top of the oscilloscope screen, and the pre-trigger portion of the waveform record is displayed to the left of it. When the trigger event takes place, the oscilloscope begins to acquire waveform samples that constitute the waveform record's post-trigger segment of the waveform record. This segment is displayed to the right of the trigger point. To ascertain the trigger event, visualize a vertical line down from the trigger point to where it intersects the displayed waveform. Following this event, the oscilloscope will not respond to other triggers until the acquisition is completed and hold-off time has expired.

In the oscilloscope front panel adjacent to the number pad are the trigger controls. They consist of one knob, which sets the triggering level, and two buttons labeled Force Trig and Menu. Pressing Menu, the triggering menu appears below the display.

The first menu option is Type. Pressing the associated soft key, we see a list of trigger types in a vertical menu on the left. The trigger type may be chosen by turning Multipurpose Knob a, which opens a horizontal menu below the display. Edge triggering is the default, by far the most commonly used. The second Trigger Type option is Sequence (B Trigger), which permits the user to select the elapsed time between A Trigger and B Trigger, by means of Multipurpose Knob a. The time range is 8.0 ns to 8 seconds.

Sequential Triggering is used when the motive is to take a detailed look at a small part of the waveform record. The user specifies the A (main) trigger event and the B (delayed) trigger event. After the A event occurs, the oscilloscope looks for a specified number of B events before triggering and displaying the more detailed segment of the wave record.

The third Trigger Type menu option is Pulse Width. When this trigger type is chosen, the oscilloscope triggers on pulses that are less than, greater than, equal to, or not equal to a specified time. Also, the user can trigger on pulse widths that are within or outside of a range of two different specified times. The user can also trigger on positive or negative pulses. Pulse width triggers are primarily used for digital signals and will be considered in more detail in Chap. 6.

The fourth menu option, Time Out, causes the oscilloscope to trigger when no pulse is detected for a specified time. The signal stays above or below a set value for that period of time.

Runt triggers on a pulse amplitude that crosses one threshold but fails to cross a second before recrossing the first. See Chap. 6 for more details on this type of triggering and also Logic, Setup and Hold, Rise and Fall Time, Video and Bus triggering. These triggering types are used primarily with digital signals.

Returning to the Triggering menu, the second menu option is Source, which may be an analog or digital channel or the AC line. Normally and by default, the trigger source is the signal that is being displayed, but it is possible to trigger from other sources, internal or external.

The third menu option is Coupling, which may be DC (the default), AC, High-Frequency Reject, Low-Frequency Reject, and Mode (Auto, which is default and Hold-Off).

Hold-Off is an interesting user-controlled option. Sometimes complex, repetitive waveforms contain intermediate peaks or other pseudo-events that cause false triggering. To prevent this from happening, the user can establish a Hold-Off time, selected by Multipurpose Knob a. This can range from the default minimum, 20.0000 ns to 8 seconds. The oscilloscope will not trigger until the next trigger event after the Hold-Off period has expired.

The fourth menu option is Slope, which permits the user to determine whether the trigger event is to occur on the rising edge, falling edge, or both edges of the waveform. The associated soft key toggles among these three choices.

The fifth menu option determines the level in volts at which the oscilloscope is to trigger, as determined by Multipurpose Knob a. This adjustment can also be made by the Level knob in the triggering controls section.

Additional triggering concepts are applicable to digital inputs only, and we will discuss them in Chap. 6.

RF

Signals may be displayed in the frequency domain by feeding them into an analog channel and pressing Math>FFT, as covered earlier. Now we'll consider an alternate method, which permits the user to access a different and more elaborate set of menu options and signal parameters.

We'll begin by running a BNC cable from AFG Out on the back panel to the RF channel input on the front panel using an RF adapter. With the AF button toggled On, press Waveform and select Sine. (It's probably already selected since it is the default.) In the AFG information bar, frequency should be 100.00 kHz, amplitude 500.00 mV$_{pp}$, and offset 0.0000 V. Any of these values can be adjusted in Waveform Settings.

Then, go to the Frequency section at the far right in the front panel, toggle On RF, and press Freq/Span. In frequency-domain displays, the vertical Frequency and Span menu is very critical if you want to see a readable display. Center frequency has to be set correctly to bring the fundamental to the center of the screen. Enter whatever frequency is in the AFG information bar (default 100.00 kHz) into the Center Frequency field, using Multipurpose Knob a or (better) the number pad. Next, if you want to display harmonics, choose an appropriate frequency such as 40.0 MHz and enter it into the Span field, using Multipurpose Knob b or the number pad. After Center Frequency and Span are entered, Start and Stop will adjust themselves. Note that R to Center has nothing to do with this operation. It sets a reference marker to the center.

When the sine wave is displayed in the frequency domain, there is one large spike, the fundamental, at the center of the screen, with no visible harmonics, as shown in Fig. 5.14.

Scrolling down to Square Wave in the Waveform menu, we see that the frequency domain display, as shown in Fig. 5.15, exhibits an array of harmonics decreasing in amplitude in proportion to their spectral distance from the fundamental.

That is because the very fast rise and fall times of the square wave are high-frequency components that create many strong harmonics.

With the square wave displayed in the frequency domain, press RF to reveal the horizontal frequency domain menu below the display. Then, push the first menu option, Spectrum Traces, to bring up the vertical Spectrum Traces menu to the right of the display. To see Average, Max Hold, and Min Hold clearly, Normal (the default) should be turned off. Then, turn Average On. This, as in the time domain, minimizes noise by averaging successive acquisitions. The number of acquisitions that are averaged can be set by the user by means of Multipurpose Knob a, and the range is from two to 512. Notice that the noise evinced by the rapidly fluctuating line along the bottom of the trace diminishes as the number of acquisitions to be averaged increases. Turn Average back to two and toggle it Off.

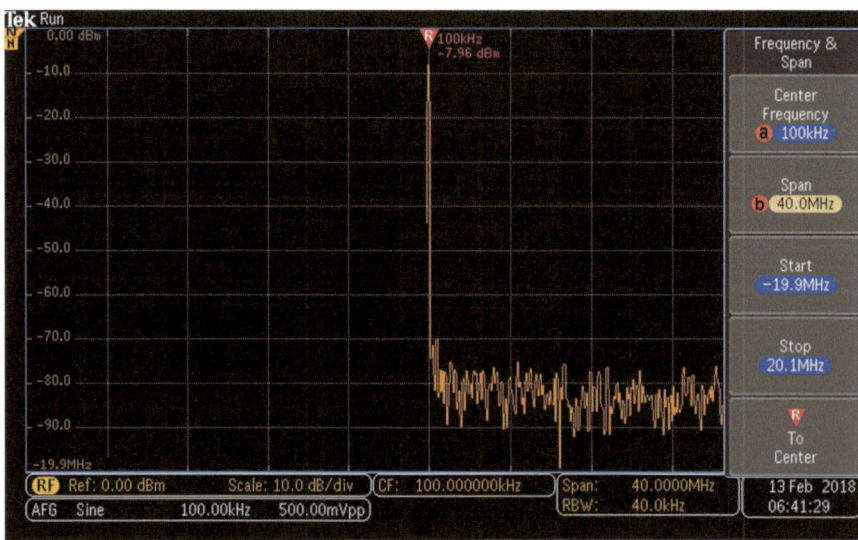

Fig. 5.14 Sine wave in frequency domain. (Author's screenshot)

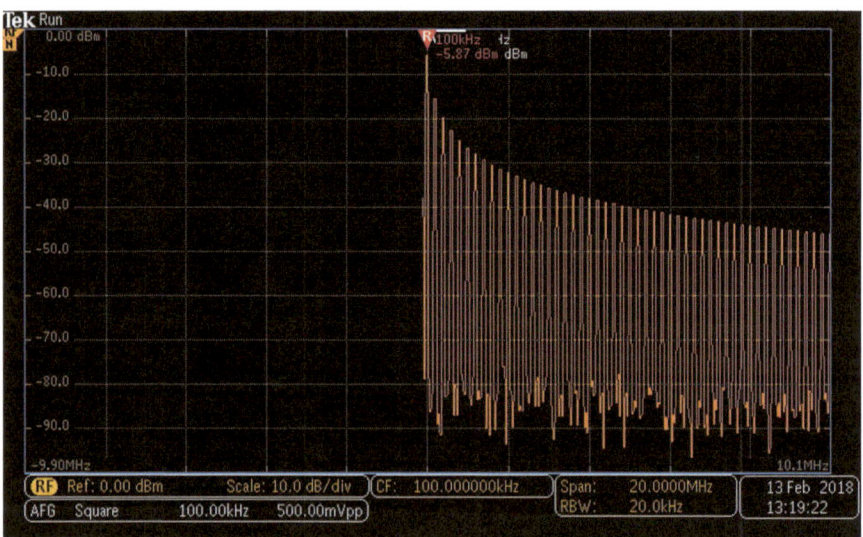

Fig. 5.15 Square wave in frequency domain. (Author's screenshot)

Toggle On Max Hold. Notice that the lower portion of the trace holds as new peaks are attained. This can be most clearly observed if you repeatedly press Reset Spectrum Traces. Min Hold is similarly interesting to watch. Be sure to turn it off when you are done. Also, turn Normal back on. Otherwise, RF won't run.

You may notice, with some signals and settings of the oscilloscope, that the upper portion of a signal, particularly where the fundamental is peaking, may be out of sight above the upper edge of the screen. You will see a warning in the lower left corner that states "Signal clipping, possible distortion." The fact is that the upper edge of the display frequently coincides with the end of linear amplification. The remedy for this situation is to press the Ampl button, just below Freq/Span. Then, in Ref Level, use Multipurpose Knob a to lower the waveform, which will also make the warning go away. You can also press the Vertical menu option and use Multipurpose Knob b to change dB/div, which resizes the waveform with respect to vertical divisions. (Pressing Auto Level may work against you.)

Utility

Returning to the time domain, press the Utility button at the bottom of the instrument. This opens the horizontal Utility menu beneath the display. Press the soft key associated with the first menu option, Utility Page. The vertical Utility Page menu appears at the left of the display. The first menu option at the top is the default Config. Of course this permits the user to set language, time and date. The fourth menu option, About, opens a window that provides the make and model of the instrument together with its bandwidth, serial number, current firmware version, and the URL where it can be downloaded with instructions for doing so and installing it in the oscilloscope. It is a good practice to watch for any new firmware version and install it when available. Also, there is a list of channels with any connected probes detected.

The final, very useful menu option is titled Manage Modules and Options. Application modules are physical units that can be inserted into the oscilloscope to provide added functionality. Afterward, the module can be removed and stored elsewhere. If the license is in the oscilloscope, the physical module need not be present. Application modules may be inserted or removed only while the oscilloscope power is off. A box on the screen lists licenses currently contained in the oscilloscope.

To summarize, the oscilloscope front panel has buttons and knobs grouped according to functions. When a button is pressed, a horizontal menu opens along the bottom and associated soft keys either toggle menu items or open a vertical submenu along the side. Menus on the left, where there are no soft keys, are navigated by means of Multipurpose Knobs a and b. Menus on the right are navigated by means of these same Multipurpose Knobs or by soft keys, which lead the user to deeper menu levels. In that manner, large amounts of oscilloscope functionality are accessed.

Reference

Horowitz, Paul and Hill, Winfield, *The Art of Electronics*, Third Edition, Cambridge University Press, 2015

Chapter 6
Acquiring, Displaying, and Measuring Digital Signals in an Oscilloscope

Abstract Using cursors to measure the value of a function. Years ago, a transparent graticule was placed over the screen of an analog CRT-type oscilloscope. The technician counted divisions and tics to estimate the amplitude and elapsed time of a trace. Greater accuracy and convenience are afforded today using the cursor function. With examples, methods for measuring values such as peak-to-peak voltage are shown. Logic gates and truth tables. Digital troubleshooting techniques.

Most wave energy in nature is analog rather than digital. The amplitude oscillations move along a continuum like the music of a slide trombone. At any instant the amplitude can be represented by one of an infinite number of values within a continuous range. A digital signal, in contrast, is composed of discrete amplitude levels located within defined bands. Not withstanding small variations, an energy level within any given band represents a unique information state.

At a certain level of scrutiny, the distinctions between analog and digital begin to evaporate. At a not quite vanishingly small spatial scale, quantum reality applies. Current thinking is that space and even time are granular. In our natural world, some birdsongs are digital, consisting of discrete tones that recur at fixed intervals and uniform volume. Some of them are even clocked and counted, separated by periods of silence.

Gottfried Leibniz (1646–1716) was one of many early theoreticians who wrote about binary numbering. In 1703 he published *Explanation of Binary Arithmetic, Which Uses Only the Characters 1 and 0*. He referenced ancient texts including the Chinese *I Ching*, showing that its hexagrams used to map abstract ideas correspond to binary numbers.

Later, the British mathematician George Boole (1815–1864) turned his attention from differential equations to algebraic logic, publishing *Laws of Thought* (1854). This revolutionary paper anticipated by a century our information age. His algebraic system of logic, which came to be known as Boolean algebra, postulated two variables, 1 and 0, which represented true and false.

As mentioned above, digital signals consist of information states corresponding to discrete values. The most common implementation is binary, consisting of two states representing logic low and logic high. There could be any number of such

D. Herres, *Oscilloscopes: A Manual for Students, Engineers, and Scientists*,
https://doi.org/10.1007/978-3-030-53885-9_6

states, and the system would still qualify as digital provided these states corresponded to discrete voltage (or current, mechanical, pneumatic, hydraulic, laser) levels (Horowitz and Hill 2015).

Binary is one of many possible digital systems, but it is the most used because it lends itself to bipolar switching formats. Early computers were analog contraptions with clunky gear trains and mechanical counters.

Nineteenth-century electrical engineers, experienced in (digital) telegraph networks, were quick to see that electricity could be used to transfer, store, and manipulate information. Since those early years, electrical technology has evolved into electronics, and increasingly these two fields have merged. Consider the programmable logic controller (PLC), where milliamp-scale information from a remote user interface controls the speed and torque of a 500-horsepower motor through the miracle of pulse-width modulation.

Logic Gates

Binary data, which has the potential to enable all sorts of physical activity, on the most basic level is processed by means of logic gates, actual and virtual. These gates in a few simple varieties can be built using electromechanical switches and relays, but today's solid-state devices are less expensive, more reliable, and use far less power.

Gates exist in a limited number of forms, but in combination highly complex architectures, even within the smallest integrated circuits, have become a reality. Binary signal processing is implemented primarily by means of these logic gates:

The AND gate, schematic shown in Fig. 6.1 and truth table shown in Fig. 6.2, is an electronic circuit that outputs a high logic state only when all its inputs are high.

Fig. 6.1 AND gate schematic symbol. (Wikipedia)

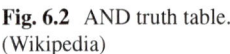

Fig. 6.2 AND truth table. (Wikipedia)

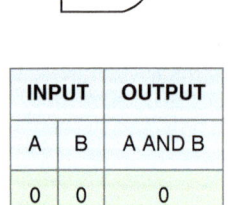

INPUT		OUTPUT
A	B	A AND B
0	0	0
0	1	0
1	0	0
1	1	1

The OR gate, schematic shown in Fig. 6.3 and truth table shown in Fig. 6.4, is an electronic circuit that outputs a high logic state only if one or more of its inputs are high.

The NOT gate, schematic shown in Fig. 6.5 and truth table shown in Fig. 6.6, is an electronic circuit that outputs an inverted version of its input.

The NAND gate, schematic shown in Fig. 6.7 and truth table shown in Fig. 6.8, is an electronic circuit that outputs a high logic level if any of the inputs are low.

The NOR gate, schematic shown in Fig. 6.9 and truth table shown in Fig. 6.10, outputs a low logic state if any of the inputs are high.

The XOR gate, schematic shown in Fig. 6.11 and truth table shown in Fig. 6.12, outputs a high logic level if either but not both of its inputs are high.

Fig. 6.3 OR gate schematic symbol. (Wikipedia)

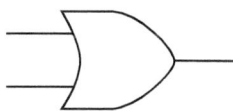

Fig. 6.4 OR truth table. (Wikipedia)

INPUT		OUTPUT
A	B	A OR B
0	0	0
0	1	1
1	0	1
1	1	1

Fig. 6.5 NOT gate schematic symbol. (Wikipedia)

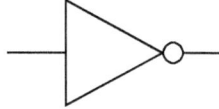

Fig. 6.6 NOT truth table. (Wikipedia)

INPUT	OUTPUT
A	NOT A
0	1
1	0

Fig. 6.7 NAND gate schematic symbol. (Wikipedia)

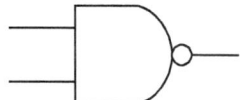

Fig. 6.8 NAND gate truth
table. (Wikipedia)

INPUT		OUTPUT
A	B	A NAND B
0	0	1
0	1	1
1	0	1
1	1	0

Fig. 6.9 NOR gate
schematic symbol.
(Wikipedia)

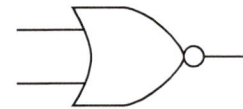

Fig. 6.10 NOR truth table.
(Wikipedia)

INPUT		OUTPUT
A	B	A NOR B
0	0	1
0	1	0
1	0	0
1	1	0

Fig. 6.11 XOR gate
schematic symbol.
(Wikipedia)

Fig. 6.12 XOR truth table.
(Wikipedia)

INPUT		OUTPUT
A	B	A XOR B
0	0	0
0	1	1
1	0	1
1	1	0

The XNOR gate, schematic shown in Fig. 6.13 and truth table shown in Fig. 6.14, outputs a low logic level if either but not both of its inputs are high.

Discrete passive and active components can be combined to make resistor-transistor logic (RTL) and diode-transistor logic (DTL) circuits, but they have fallen out of favor, replaced first by transistor-transistor logic (TTL) and currently by CMOS-type MOSFET logic, which is characterized by higher-speed and lower-power dissipation.

Viable gates can be built using electromechanical components such as switches and relays. These devices can be configured to create inputs and outputs that reflect valid truth tables, though at greater expense and less reliability than our user-friendly semiconductors. There is a more fundamental difference: the electromechanical switch constitutes a continuous conductive path so that current can flow in either direction between input and output. Current does not flow between the input and output of a semiconductor-based logic gate. The semiconductor logic gate, more-over, draws a minute amount of current at its input, while the output is a stiff low-impedance source that is capable of enduring a range of loading conditions.

Nevertheless, there are limits to the number of gates that can be cascaded. Each logic gate, like any semiconductor, has an intrinsic propagation delay, the time required for the output to respond to any changes applied to the inputs. Naturally this propagation delay is cumulative when multiple logic gates are wired in series.

The NOT gate, as we have seen, has one input. Other logic gates have two or more inputs. Except in a few specialized applications, there is only a single output. (After all, it wouldn't make sense to have two or more differing outputs, and if they were the same, that simply equates to redundant pins).

When digital logic gates are connected making higher-level logic circuits, unused inputs at the gate level must be connected to either a logic high or logic low level, using a reasonably sized pull-up or pull-down resistor to stabilize the voltage. Otherwise the unused input would be subject to floating potential, producing unpredictable switching.

Fig. 6.13 XNOR gate schematic symbol. (Wikipedia)

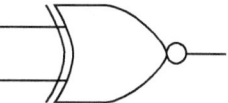

Fig. 6.14 XNOR truth table. (Wikipedia)

INPUT		OUTPUT
A	B	A XNOR B
0	0	1
0	1	0
1	0	0
1	1	1

An individual logic gate embodies combinational as opposed to sequential logic. A logic gate is intelligent and absolutely precise, but it has no memory or knowledge of any past state. A single logic gate cannot store data, its output relating only to the current state of its inputs. Data storage enters the picture when at least two logic gates are configured to create a latch circuit. This device is bistable, meaning that the inputs can cause the output to assume either a logic high or a logic low state. The latch will remain in that state even after the inputs are deactivated. A latch, then, is a single unit of memory. It embodies sequential logic.

Latch and Flip-Flop

Some texts use the terms latch and flip-flop interchangeably. While both are sequential bistable multivibrators and qualify as memory devices, in current usage flip-flop refers to a clocked circuit. According to that distinction, the latch is level sensitive, whereas the flip-flop is edge sensitive. A latch is sometimes called a simple flip-flop, as shown in Fig. 6.15. This device can be constructed using two cross-coupled inverting devices: vacuum tubes, bipolar transistors, field-effect transistors, inverters, or logic gates.

Digital circuit analysis, as required in troubleshooting malfunctioning electronic equipment and in debugging new designs such as prototypes created for product development, consists primarily in looking at outputs. When they do not appear abnormal and yet the equipment is subject to crashes or other unexplained behavior, the cause may be infrequent transients, which are notoriously difficult to see. The oscilloscope, in conjunction with specialized setups and a measure of expertise, provides a window through which infrequent and often random glitches can be located and related to other occurrences such as events in the analog sector, notably power supply anomalies. We'll get to that later in this chapter, but to begin we'll cover some very simple procedures.

Fig. 6.15 A simple non-clocked flip-flop based on two cross-connected bipolar junction transistors. (Wikipedia)

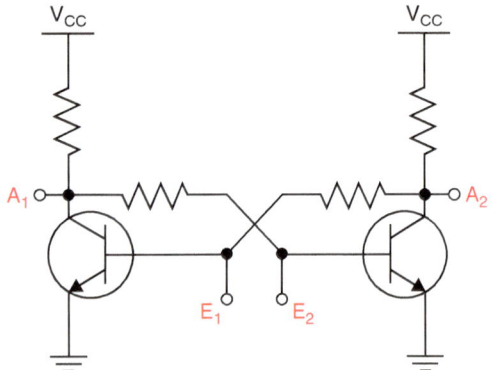

It is sometimes supposed that test equipment intended for electrical work is useless for analyzing digital equipment. This is not true. The simple multimeter is a little more time-consuming to use compared to the logic probe, but it can provide valid and in some applications more extensive results.

Multimeters range in cost and quality from a $10 meter that can be bought in a big-box store to a $500 and up instrument that will last a lifetime and has many more features and capabilities. All multimeters measure volts, milliamps, and ohms, but upscale models also offer diode and capacitor test and continuity check with an audible beep at under 30 ohms. A good model has clamps on the back side that can hold one or both probes so that you don't need three hands to contact two points on a circuit board and simultaneously hold the meter. The ohms mode is invaluable for checking circuit-board traces, terminations, and solder joints for invisible breaks.

The logic probe is a very useful tool for digital troubleshooting and diagnosis. This is a handheld probe with a conductive needlepoint designed to contact closely spaced non-insulated conductors, circuit nodes, and terminations on a circuit board. Most of these instruments have three LED's on the probe body. The red LED indicates high logic state. The green LED indicates low logic state. An amber LED indicates a pulse. A pulse stretcher circuit enables the amber LED to denote very brief pulses. A control on the probe body enables the user to place the instrument in single-event storage or free-running mode. When the logic probe is connected to an invalid or non-energized source or not connected, none of the LED's are on. A slide switch on the probe body permits the user to place the instrument in either TTL or CMOS logic mode, with their differing voltages.

A valuable feature in some probes is an audible tone indicator, the higher-pitched tone denoting logic high. This is desirable when checking dense areas where it is difficult to probe and watch the LED's simultaneously.

The value in a logic probe is that it permits the user to jump around very quickly and accurately throughout complex circuitry so as to get an idea of what is going on. Another inexpensive tool is the logic pulser, spelled with an e. (A pulsar is a very rapidly spinning neutron star that like a rotating beacon in a lighthouse appears to flash when its beam strikes earth.) The logic pulser injects CMOS or TTL logic high or low pulses into the wire or terminal that is probed. It is needed when there is no signal at the inputs due to dead upstream circuitry. While less frequently used than the logic probe, it is sometimes essential.

A logic analyzer, rack-mounted, bench-type, or portable, is conceptually similar to the logic probe but with far more extensive features. Combined with advanced user-determined triggering capabilities and the ability to display multiple channels, it may convert data into timing diagrams, protocol decodes, and other formats.

A mixed-signal oscilloscope (MSO) or mixed-domain (MDO) oscilloscope with MSO-enabled protocol analyzer combines a portion of the logic analyzer's functionality with an ability to see analog and digital signals simultaneously, in either time or frequency domain.

Conveying Digital Information Point to Point

We've seen how analog signals are converted to digital signals by means of sampling in the ADC. The new digital signal carries the full amount of information that the analog signal contains provided the sampling rate complies with the Nyquist-Shannon requirement. (As a practical matter, sampling is usually implemented at a somewhat higher rate to ensure a seamless transition.)

But after analog-to-digital conversion has been accomplished, the new digital signals have no value to anyone unless they can move to a memory, processor, or real-time display equipment. In the old analog oscilloscope epoch, the analog signal, following amplification, attenuation, and filtering, was simply applied to the vertical deflection plates in the CRT, where in conjunction with a precisely triggered ramp signal applied to the horizontal deflection plates so as to create a time base, a trace appeared on the screen. (We have to remember that this trace was not the waveform but, properly speaking, a graph of its amplitude with respect to time drawn in reference to Cartesian Coordinates.) (Gibilisco 2012)

Digital signal transmission is a little more complex. It can take either of two forms – parallel transmission or serial transmission. The older method, parallel transmission, is easy to understand. Multiple data bits, usually in 8 bits, single-byte units, are conveyed simultaneously on bundled, insulated wires or separate circuit-board traces, or, also simultaneously, via separate but closely related frequency channels. Parallel data transmission is conceptually simple, but, like many old-world technologies, there are inherent inefficiencies and vulnerabilities. It is more prone to crosstalk, noise, and skewing (data corruption due to slightly different conductor lengths) and more expensive to build and maintain with all those wires that have to be terminated at both ends.

Bus

Serial data transmission is a higher tech and cleaner protocol. It exists in quite a number of radically different formats, each requiring a different body of knowledge and expertise. We'll do a rundown of the principle varieties, but first we'll have to define "bus."

Originally derived from the Latin word *omnibus*, which means "for all," bus as used by electricians refers to a short conductor, frequently rectangular in cross section and mounted inside an enclosure, that is designed to carry high current at substantial voltage relative to ground.

In digital signal transmission, bus refers to the transmission media, not restricted to the conductors but including terminations, ancillary equipment such as impedance-matching devices, protocols, installation procedures, and in fact anything relating to the link between transmitter and receiver. It also includes media inside the equipment, such as circuit-board traces that may convey digital signals between integrated circuits and short lines from one circuit board to another. Digital bus is a very broad term, and there are numerous varieties, with various working voltages, protocols, and encoding schemes. Here's a rundown of the principle types:

I²C

I²C denotes inter-integrated circuit. (The superscript is a convenient abbreviation that has no mathematical meaning.) As a bus, I²C most frequently appears as traces on a circuit board or short conductors between circuit boards within the same equipment, facilitating digital communication between microprocessors.

Siemens AG, NEC, Texas Instruments, ST Microelectronics, Motorola, and others developed compatible I²C products, which are now widely used. There are no licensing fees.

I²C is a two-wire bus consisting of any number of masters and slaves on a single digital bus. These individual devices are connected in parallel across the two lines. Initially, logic high is connected to one of the lines, and it is pulled down in the event that either a master or a slave conducts. In that way, a single device can impose a digital signal on the bus.

Both conductors are bidirectional, open-drain lines. One is a serial clock line, and the other is a serial data line. A node may be master or slave depending upon its function.

A slave, continuously receiving the clock signal, responds when its unique address is sent by the master. At the end of any given transmission, master and slave may exchange identities.

The 7-bit address that identifies a particular slave is followed by a 0 if the master wants to write or by a 1 if the master wants to read. If both lines are high, the bus is idle. When a device conducts so as to drive the bus low, the system becomes active.

The I²C bus is monitored for start and stop bits by all masters. Only one master at a given time uses the bus, and during this event, no other master will transmit. If two masters happen to initiate communication simultaneously, the one speaking to a slave with the lower address will have priority.

Accessing an I²C Signal

To demonstrate what an I²C signal actually looks like, Tektronix provides a proprietary Demo One board, with dedicated terminals for accessing a number of digital bus signals. The board is powered by the oscilloscope through a specialized Y-type USB cable that has two parallel inputs. They are to be plugged into separate USB slots in the oscilloscope. This USB cable carries power only, no data. The two separate inputs are needed because the board draws a significant amount of current. If you plug in just one branch, you can overload the oscilloscope's USB output. The hookup is shown in Fig. 6.16.

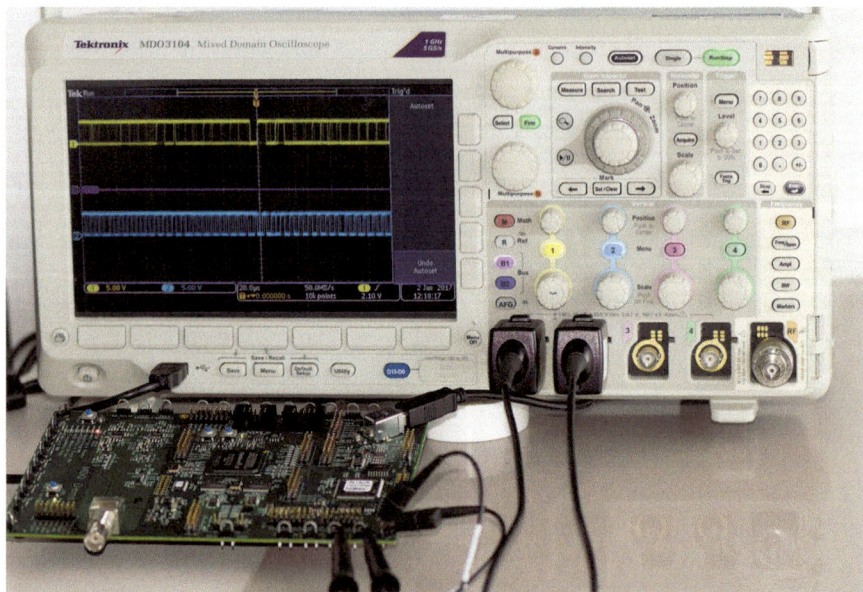

Fig. 6.16 A Tektronix MDO3000 series oscilloscope accesses an I²C signal from a Demo One board. (Tektronix)

The Demo board is actually designed to go with the MSO4000 oscilloscope, but it works well with other instruments provided current requirements are observed. One red and two green LED's indicate the board is powered up and ready for use. With 10:1 probes plugged into two analog channels, both ground return leads are connected to ground terminals, and the hook-type probe tips are connected to I²C clock and data terminals. The signals are displayed as shown in Fig. 6.17.

The analog channels can be scaled to display simultaneous clock and data signals, and the acquisition can be stopped or zoomed and panned using Wave Inspector. Having seen I²C bus clock and data signals in a controlled context, we are ready to look at signals from functioning or non-functioning equipment for the purpose of troubleshooting or design.

A more detailed analysis of the I²C bus can be performed using bus trigger, decode, and search tools. Here are the procedures:

- Connect Channel One probe to the I²C_Clk signal on the Demo One board.
- Connect Channel One ground return lead to the nearest GND terminal.
- Connect Channel Two probe to the I²C_Data Signal on the Demo One board.
- Connect Channel Two ground return lead to the nearest GND terminal.
- Press "Recall Demo Setup."
- Press the B1 button in the Bus section, then Event Table lower menu button, and then Event Table right menu button to turn on the decode Event table.
- Press Previous (<) and Next (>) front panel button to navigate through the search marks showing all "Address 50" events in the record.

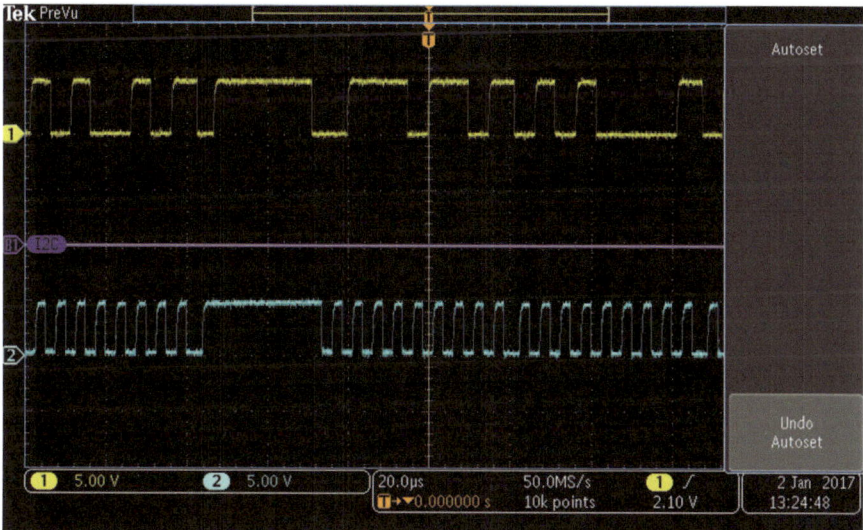

Fig. 6.17 I²C signals displayed in Tektronix MDO3000 series oscilloscope. (Tektronix)

The I²C bus is used in most embedded systems designs for chip-to-chip communication and configuration of components. In the past, technicians and researchers have had to decipher the bus manually, counting bit numbers. With the optional embedded application module, the Tektronix MDO3000 series instruments can perform the decode manually, which is less time-consuming and more accurate. The instrument can also trigger or automatically search packet level information, focusing on packet contents such as address values, data values, missing acknowledgment, and the like.

RS-232 Bus

As shown in Fig. 6.18, RS-232 is an old technology, largely eclipsed by the newer USB hookups. We have all seen the old trapezoidal RS-232 terminals in our first PC's. This old parallel bus, however, is still going strong in scientific instrumentation and on the factory floor where it links programmable logic controllers (PLC's) to high-horsepower motors and all kinds of actuators. If you ever travel to the International Space Station, you'll see it used to good effect there, where reliability is a primary consideration. No doubt you'll connect RS-232 conductors to the logic analyzer that is built into a contemporary oscilloscope in order to check timing relations and signal integrity.

In addition to scientific instrumentation, RS-232 bus hookups in production facilities generally terminate at user interfaces or, during initial setup, at ruggedized laptops. Besides RS-232, RS-422, RS-485, or Ethernet may be used. Programming

Fig. 6.18 A 25-pin connector used to terminate the RS-232 bus. (Wikipedia)

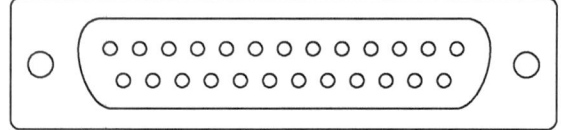

software takes various forms. Ladder logic is common because factory-floor electricians can easily use it for new setups and quick alterations. RS-232 terminations work well in harsh factory environments characterized by abrupt temperature changes, dust, vibration, and moisture.

RS-232 may consist of synchronous or asynchronous transmission of data and control signals. The individual circuits work in one direction only, from computer or CPU to peripheral, or on separate wires in the other direction. Because separate parallel connected pairs operate simultaneously, full duplex information transfer occurs. Character framing and encoding as in serial buses is not used.

RS-232 voltage levels are +3 to +15 for logic high and −3 to −15 for logic low. The restricted zone, −3 to +3 with respect to a common ground pin, is not valid. RS-232 has to see a stable zero potential on the ground pins at both ends of the media. Otherwise, there will be circulating electrical current. Accordingly, RS-232 exists in short lengths only. RS-422 and RD-485 employ balanced transmission with differential signaling and common mode rejection, so longer cable lengths are feasible.

Often there are far fewer conductors than the maximum permitted by termination hardware. Postal meters and GPS receivers that transmit position only have two wires, data and ground, while five conductors are needed for two-way data along with hardware control.

A digital probe and pulser will get you started in troubleshooting an RS-232 bus, but for a more complete sense of what is going on, you will want to bring out an oscilloscope. To begin, check individual pulses in an analog channel to make sure that voltages are good for the logic levels and that the pulses are not overly noisy as is sometimes the case in parallel digital transmission. For comparison, an RS-232 transmission signal, shown in Fig. 6.19, is available in the Tektronix Demo One board. The oscilloscope probe tip is connected to the labeled terminal, and the ground return lead is connected to the nearest GND terminal.

Serial Peripheral Interface (SPI) Bus

The serial perimeter interface (SPI) bus resembles the I²C bus. Both consists of masters and slaves, which interact with one another, but beyond that there are significant differences, particularly in their wiring, as shown in Fig. 6.20.

The lines consist of master-out, slave-in (MOSI), which conveys data from the master to whatever slave is active. Inactive slaves disregard it.

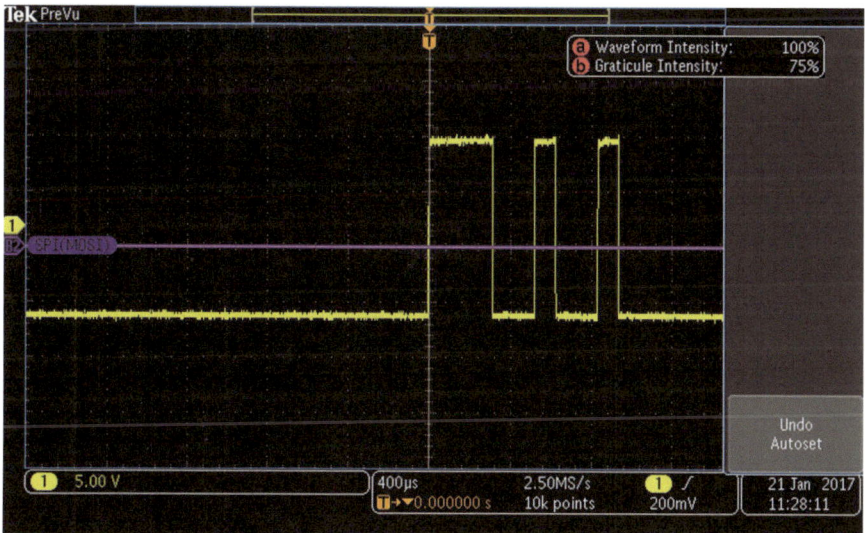

Fig. 6.19 RS-232 transmission signal. (Tektronix)

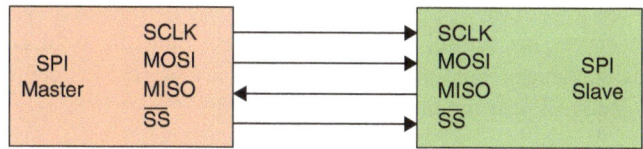

Fig. 6.20 Most SPI installations consist of a single master device and multiple slaves, although occasionally there is only one slave. (Tektronix)

The other line is master-in, slave-out (MISO), which conveys data from an active slave to the master. Data can be transferred by these separate lines simultaneously. Individual serial clock lines connect the master to each slave. The clock, in the master, outputs a square wave to the active slave. The SPI bus user determines the clock frequency, which cannot be faster than permitted by the slowest slave.

Slave select (SS) consists of separate lines that go to each slave. At start-up, a logic high voltage is conveyed to all slaves. The master selects a slave by switching to logic low for that slave. Thus, unlike in I²C, slave addresses are not needed.

The Tektronix Demo One board has terminals for SPI clock, MISO, and MOSI signals, in addition to ground terminals for all lines. An SPI display is shown in Fig. 6.21.

A fourth oscilloscope channel could be used to display an SS signal, but since this is merely logic low or logic high, a multimeter will suffice.

SPI is used for short-haul transmission, and it figures prominently in embedded systems. You'll see it in secure digital cards and liquid crystal displays. It is not to be confused with synchronous serial interface (SSI), which employs differential signaling and one simplex channel.

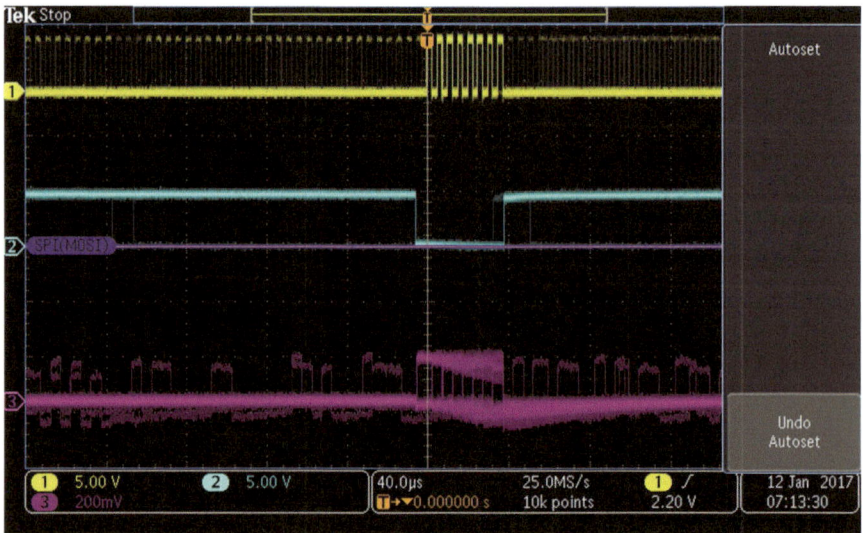

Fig. 6.21 Three oscilloscope analog channels and probes are required to display the SPI bus in a Tektronix MDO3000 series oscilloscope. (Tektronix)

SPI has been very successful where board space is limited and the environment offers unique challenges. For example, you'll find it in digital imaging equipment such as the Canon EF lens mount and some SD cards.

USB Bus

USB is a serial bus that provides a data and power connection between a computer or suitable USB host and peripherals, storage media, and numerous compatible devices.

Beginning in 1994, a number of consortiums have promulgated a series of versions, the most current being USB 3.2 released as a specification in September 2017.

A USB system consists of a single host with downstream ports and peripherals connected in a star topography. Host devices may have two or more ports. Optional USB hubs permit the connection of up to 127 devices. The host's internal hub is known as the root hub.

The most visible part of the USB hardware is the cable with plugs on either end. This comprises the media between host and device. Additionally, if the device is not separately AC-powered, the host may power the device through the cable. Another scenario is that the device may have batteries that are charged through the USB cable.

Most USB cable ends are not interchangeable. The host end is A-type, and the device end is B-type. The A-types (with variations) are mostly the same and work cross-platform. USB connectors come in three sizes: standard, for desktop and portable devices, mini for mobile devices, and a thinner micro for slim mobile equipment such as cell phones and tablets.

USB cords and terminations are rugged and trouble-free. Cords are less expensive than hosts and devices, so by design the plugs are subject to more wear, sparing the ports. Most terminations will withstand 15,000 insertions.

USB has a specialized, almost arcane terminology. The connectivity between host and device is made up of "pipes." These are conceptual elements, not hollow cylindrical objects. These pipes carry information between the host and "endpoints" which are located in the device. Up to 32 endpoints (16 in and 16 out) are assigned numbers during initialization at start-up.

The USB protocol provides for two kinds of pipe, message and stream. A message pipe conveys brief commands from the host to the device, plus responses in the reverse direction. Stream pipes are one-way. They are arranged to transmit data between host and device. At the start of a data transfer, the host sends a token packet stating the destination, which is specified by an address and endpoint number.

As with certain other bus types, signals, as shown in Fig. 6.22, are available in the Tektronix Demo One board.

Begin by connecting analog Channel One to the USB FS signal pin with the arrow on the J601 square pin on the Demo board. Then connect Analog Channel Two to the USB FS signal pin opposite the arrow on the J601 square pin. Connect ground return leads of the probes to the nearby GND terminals on the Demo One board. Press "Recall Demo Setup." Then, in the bus section on the oscilloscope

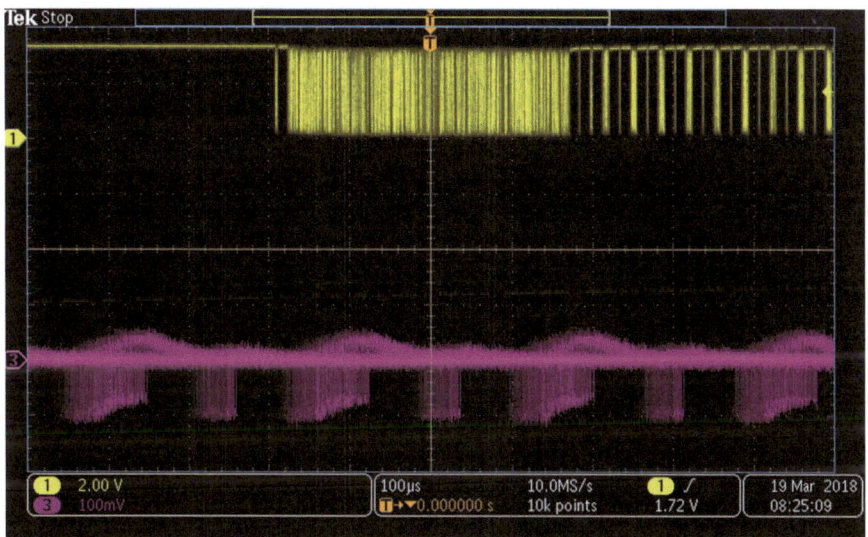

Fig. 6.22 The demo shows USB bus analysis using serial bus trigger, decode and search tools. (Tektronix)

front panel, press the B1 button, the Event Table lower menu button, and the Event Table side menu button to turn on the Decode Event Table. Press Previous (<) and Next (>) front panel buttons to navigate through the search marks showing all OUT token packet events in the record.

MIL-STD-1553 Bus

MIL-STD-1553 is a digital serial bus, first used by the US Air Force in the F-16 Fighter to connect electronic devices to a processing unit and human interface. The bus quickly spread to other avionics and spacecraft applications. It is currently used in aircraft throughout the world including the Russian MIG-35.

Reliability is implemented in a robust physical layer incorporating multiple redundant elements. The bus media consist of 75-ohm conductor pairs operating at 1 MHz. Transmitters and receivers couple into the bus through isolation transformers, limiting the possibility of a fault to chassis ground.

The bus is doubly or triply redundant due to independent wire pairs. All devices are connected to all buses. In each bus is a single bus controller and multiple remote terminals. Additionally, bus monitors record all activity in case later analysis becomes necessary, but they do not play an active role in data transfer.

The bus controller and all connected remote terminals have access to all transmissions. Communication is always initiated by the bus controller, which can permit communication between terminal devices.

Remote terminals facilitate the great number of functions that are required to make a large aircraft fly. This includes everything from engine performance to landing lights and doors.

As for cabling, MIL-STD-1553 states only that it is to be shielded. Manufacturers use twinax, which resembles coax but has two inner conductors instead of one. The insulated pairs are balanced and twisted to limit crosstalk and electromagnetic noise. The dielectric filler material minimizes capacitive coupling. The outer jacket has a high thermal rating.

MIL-STD-1553 bus signals, shown in Fig. 6.23, are available in the Tektronix Demo One board.

To access the MIL-STD-1553 serial bus in the Tektronix Demo One board, connect analog Channel One to the MIL-STD-1553+ signal on the Demo Board J1009 header. Connect the probe ground return lead to a nearby GND terminal. Then press "Recall Demo Setup" in the bus control section on the oscilloscope front panel and press the Event Table lower menu button followed by the Event Table right menu button to turn on the Decode Event table.

Press the Previous (<) and Next (>) front panel buttons to navigate through the search marks showing all sync events in the record.

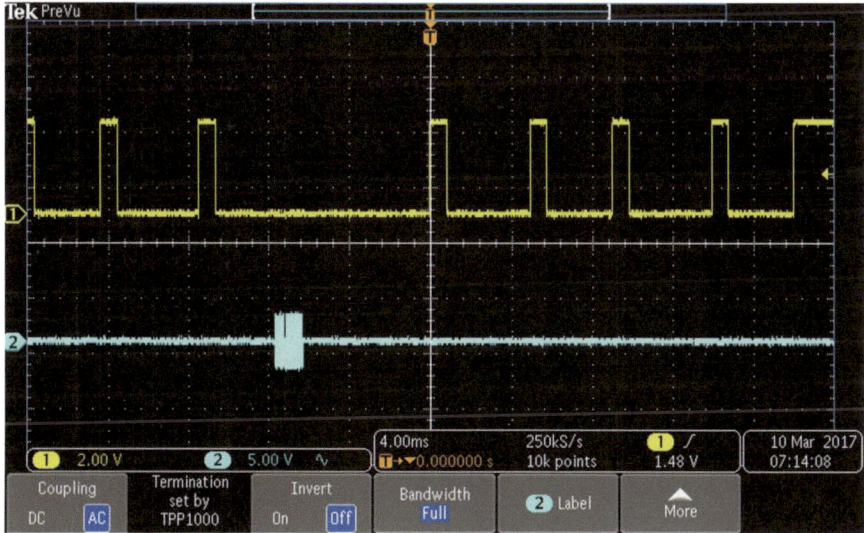

Fig. 6.23 A Tektronix MDO3104 oscilloscope displays MIL-STD-1553 bus protocol traffic running at 1 MHz. Messages consist of one or more 16-bit words, each of which is preceded by a 3 µs sync pulse followed by an odd parity bit. (Tektronix)

CAN Bus

Controller area network (CAN) bus currently dominates digital connectivity in automotive equipment, where it is best known. But following introduction in 1983 by Robert Bosch GmbH, a multinational engineering and electronics giant headquartered in Germany, CAN bus has permeated numerous fields such as elevator electronics, where low cost and extreme reliability are priorities. It is also used to good effect in video games, human prosthetics, and highly technical areas such as particle accelerators, to collect data on subatomic particle collisions.

Individual nodes in a CAN bus system receive all transmissions except where filters are incorporated to reject irrelevant messages. AND logic is used throughout. Accordingly, a node can impose a logic zero state on the CAN bus regardless of what any individual node may say.

CAN bus typically conforms to the two-wire balanced signaling format, which is called high-speed CAN. It takes advantage of differential signaling, characterized by fault tolerance and noise immunity, which is necessary in automotive applications. Common mode rejection reduces noise by means of balanced differential signaling in shielded or unshielded twisted-pair configuration. A high twist rate facilitates capacitive and inductive decoupling, always a plus from the point of view of signal integrity. Among serial buses, CAN bus is highly successful, due to its simplicity and easy implementation, despite which it performs complex tasks in extraordinarily demanding environments.

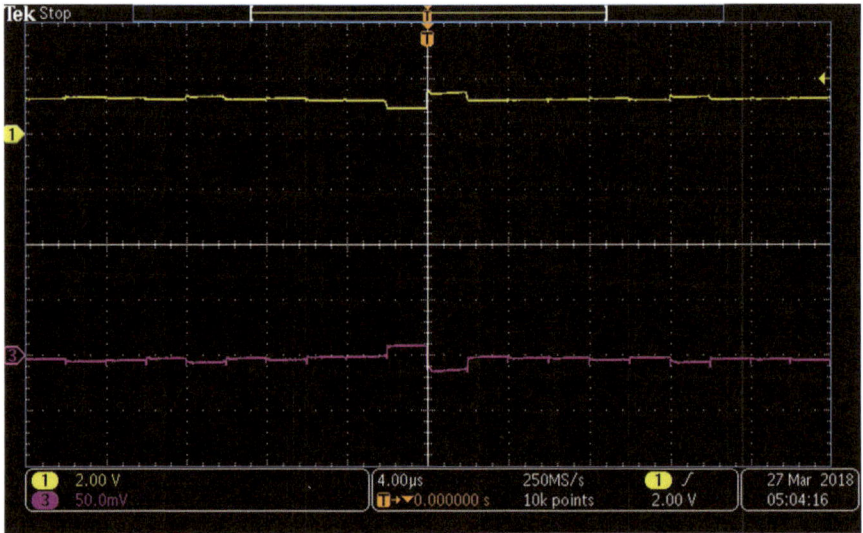

Fig. 6.24 High-speed CAN Bus. (Tektronix)

Figure 6.24 shows a high-speed CAN bus signal. Note that the two lines comprise a differential pair. This suggests a troubleshooting strategy. If one of the lines doesn't look right, try reversing the connections at both ends of the transmission line. This will reveal a line fault if that is the problem.

In the Tektronix MDO3104 oscilloscope, pressing the Utility button adjacent to the Save/Recall section below the display, a horizontal menu appears. Press the soft key associated with Utility Page and then use Multipurpose Knob a to navigate to Demo. Then press the soft key associated with Serial Bus, and in the vertical menu on the right, press the soft key below CAN to see a CAN bus analysis of serial bus trigger, decode and search tools. Connect Channel One to the CAN signal on the Demo One board. Connect the Channel One probe ground return lead to a nearby GND terminal.

Press "Recall Demo Setup" and then press the B1 button and Event Table lower menu button to turn on the Event Table. Press Previous (<) and Next (>) front panel buttons to navigate through the search marks showing all identifier x519 events in the record.

Local Interconnect Network (LIN) Bus

CAN bus, LIN Bus, and FlexRay bus are related technologies that are combined in new automotive applications in the interest of economy without a quality trade-off. That is possible because the three buses are implemented in separate areas that are more and less critical from the standpoint of safety.

Following the widespread adoption of CAN bus, first by Mercedes Benz in 1991, and soon throughout the automotive world, engineers wanted a less expensive bus for communication among less critical car parts such as power windows, mirrors, doors, and the like. LIN Bus was created to meet the need. It consisted of a lower bandwidth multiplexed bus with less costly 8-bit microcontrollers. The format was an embedded serial universal asynchronous receiver/transmitter (UART).

A single master and one or more slaves comprise LIN Bus. This comparatively frugal protocol results in a less expensive bus that consumes less power. It is possible because the LIN Bus devices enter a sleep state when they are not needed. They do so when the master sends a request frame whose first data byte is zero. Also, when the bus has been inactive for 4 seconds, the devices time out. A user, moreover, can induce sleep by setting the LIN sleep attribute to True.

The LIN Bus can operate on 12 VDC, making it suitable for automotive use without power conversion.

LIN Bus operates alongside CAN bus in fault-tolerant applications that are not time critical, which simplifies the hardware. Master-slave communication is one-wire with chassis ground. Maximum conductor length is more than 130 feet, which is enough to work in cars and trucks.

Costs are minimized in many ways, such as using RC (rather than quartz) oscillators for local clocks. With the economic incentives, manufacturers have embraced LIN Bus, shown in Fig. 6.25, and it appears to be here to stay.

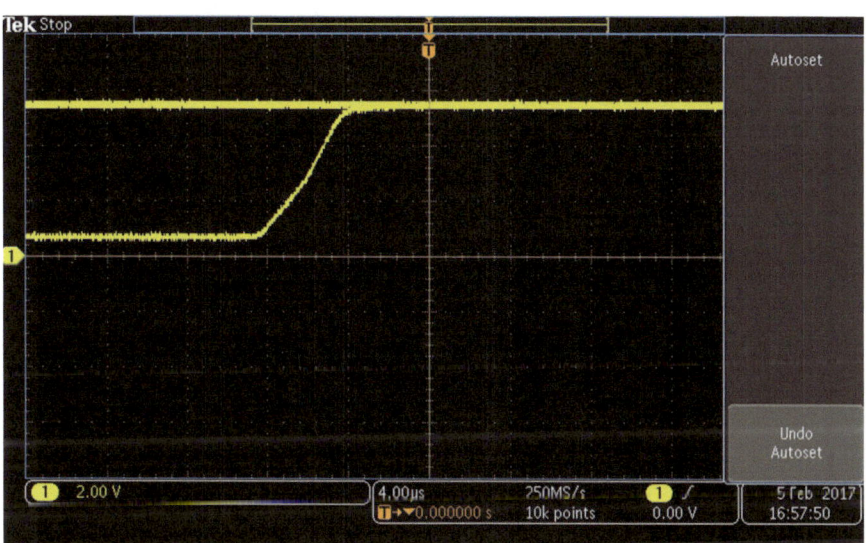

Fig. 6.25 In the Tektronix Demo One board, the LIN Bus signal is located at grid location A8. It is made up of the bus signal between two LIN transceivers. The bus speed is 19.2 kbaud. It contains a mix of version 1 and version 2 frames. The slow voltage transition translates to a reduced bandwidth requirement. (Tektronix)

FlexRay Bus

We've seen how LIN Bus complements Can bus by providing a low-tech alternative that can be used in less critical applications. FlexRay bus is the other side of the coin. It is a high-tech alternative that is appropriate in the most critical applications. All three of these buses can be combined in a single automobile or industrial unit, and they will function autonomously without affecting one another.

In contrast to LIN Bus, which dispenses with collision detection, uses one wire for lower-bandwidth communication, and has simpler components to further reduce cost, FlexRay bus provides premium connectivity for the most critical applications. It is ten times faster than CAN Bus. Where LIN Bus propagates on one wire and CAN bus on two wires, FlexBus has the option of going on either two or four wires, depending upon the task.

In new automobiles, FlexRay Bus, shown in Fig. 6.26, is used for the high-performance drive train, drive by wire, active suspension, and adaptive cruise control.

FlexRay bus cabling is one or two differential twisted pairs, with power and ground lines run independently to all nodes. Enhanced fault tolerance and increased bandwidth combine to achieve high-speed, reliable performance. Data reflections, collisions, and loss are prevented by 80- and 110-ohm resistors placed at the end nodes. In this way, multiple drops are protected.

CAN Bus, LIN Bus, and FlexRay bus have elements in common. All of them often have multi-drop topologies. But FlexRay bus typically uses more advanced star topologies, resulting in enhanced reliability and performance.

Borrowing from CAN bus and LIN Bus, multiple engine control units can be connected to a single FlexRay bus in multi-drop topology. But more advanced FlexRay designs use the star network, which is preferable. This configuration resembles a computer's LAN Ethernet implementation. Star topology is better because longer spans are possible since wire doesn't have to be run from node to node. It's not just a matter of maximum length. Shorter spans are more economical, and there is less exposure to physical damage.

Debugging and Troubleshooting Digital Circuits

Embedded systems and stand-alone digital electronic equipment consist of an array of electronic components and circuits. Typically, there are IC's mounted on printed circuit boards with signal and power inputs and also outputs ultimately connected to LED's, speakers, flat-screen displays, motors, or any of a variety of devices that are

Fig. 6.26 FlexRay two-channel cabling. (Tektronix)

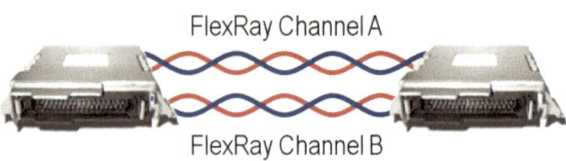

FlexRay Channel A

FlexRay Channel B

known as appliances. Another configuration is that the output is conveyed by cable or wirelessly to become the input of another stand-alone unit, as in a component audio system. Some of these transmitted signals are analog, while others are digital. In this chapter we are focusing on the digital signals.

In the oscilloscope's protocol analyzer section, we'll be looking at as many as 16 channels, so depending on the number of lines in the buses, a varying number of signals can be displayed. And even in an MDO instrument, provided the protocol analyzer is MSO capable, as many as four analog signals in addition to the digital signals can be displayed in split-screen format. All signals are represented on a single time base.

The best way to begin is to apply the digital signals to analog inputs and make sure that the information they contain is suitable for the creation of high and low logic levels. A certain amount of noise is inevitable, and it will not be present in the digital transmission provided amplitude variations do not compromise the formation of high and low logic levels.

In the digital domain, these pulses are seen as digital signals. They are either logic high or logic low. You won't see the detail that is displayed in the analog time domain or in the analog frequency domain.

In analog channels and in the digital channels of the protocol analyzer, everything is shown on a single time base. Of course you can modify the representation of this time base by adjusting the scale knob in the horizontal section, but that is just equivalent to changing the time per horizontal division in the display.

In today's digital world, signals of ever-higher frequency are the norm, which enhances speed and performance. But when it comes to debugging and troubleshooting, this increased speed can be problematic. It is becoming a massive undertaking to look through miles of periodic waveforms to find a glitch or runt (to be defined below).

The Wave Inspector section in the middle area of the front panel consists of two concentric knobs and eight buttons that permit the user to find, isolate, and measure what are known as infrequent anomalies. Figure 6.27 shows a sequence of pulses that are too close to determine if hidden among them are infrequent anomalies that could cause equipment malfunction.

Figure 6.28 shows the same pulse with Wave Inspector turned on.

The bracketed portion of the original waveform appears in the bottom portion of the display. The button with the magnifying glass icon in the Wave Inspector section toggles that facility on and off.

The smaller inner concentric knob zooms the waveform in the lower window in and out. This is reflected in the upper window where the brackets move closer and farther apart. The larger outer concentric knob pans forward and backward through the displayed waveform. This is reflected in the upper window by the brackets moving from side to side.

The button with the start-stop icon causes the Wave Inspector to automatically pan through the waveform. When it reaches the end of its travel, the movie stops. It can be repositioned manually, using the pan knob. Using the three marker buttons in conjunction with pan, multiple marks (white triangles) can be set and cleared. If you see an anomaly or point of interest, you can mark it for later study.

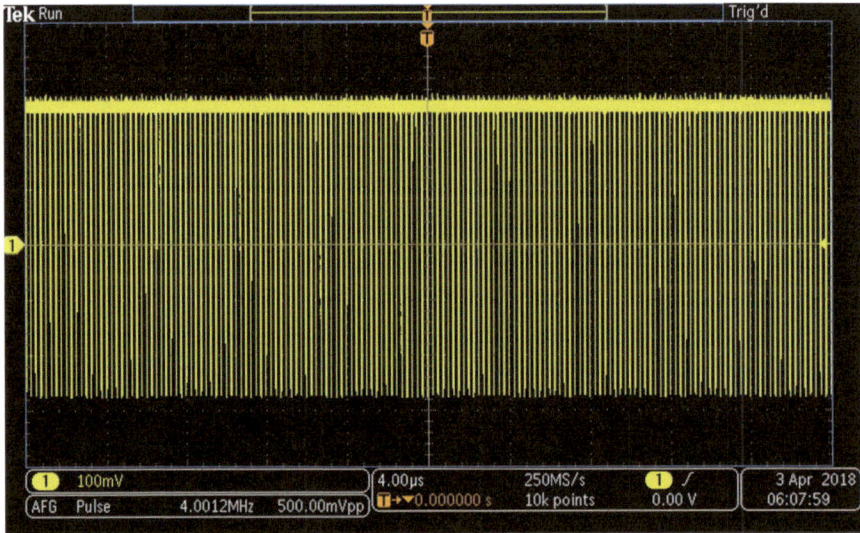

Fig. 6.27 A 4 MHz pulse signal from the internal AFG. (Tektronix)

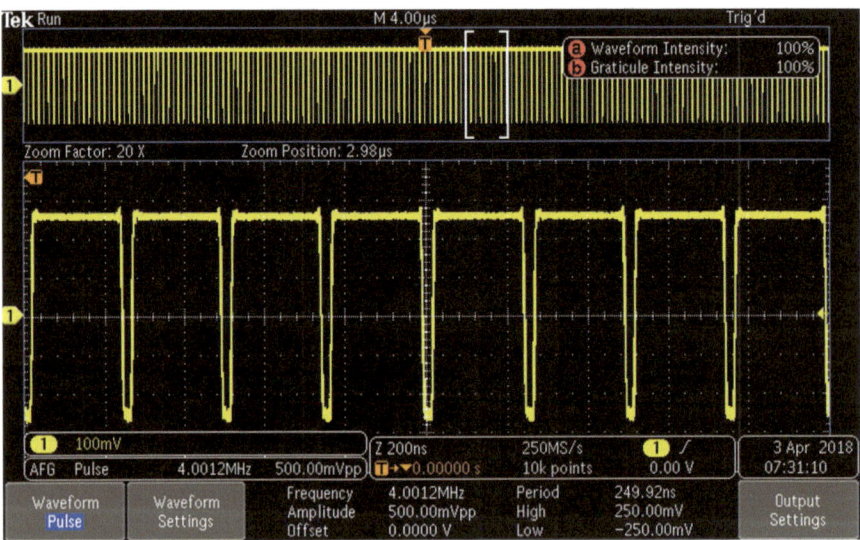

Fig. 6.28 Original signal and Wave Inspector trace in split-screen format. (Tektronix)

The Search button permits Wave Inspector to place markers at user-selected locations. The Search button toggles the facility on and off. When it is turned on, horizontal and vertical Search menus appear at the bottom and right side of the display. The first thing to do is to turn on Search in the vertical menu. This fully activates both menus.

In the horizontal menu, Search Type permits the user to set parameters using Multipurpose Knob a for edge, pulse width, time-out, runt, logic, setup and hold, rise-fall time, and bus. Each of these changes one or more of the menu selections in the horizontal menu. Then, each of those selections open an array of options in the vertical menu.

In the Wave Inspector section, Measure opens a horizontal menu. Measurements can be added or removed, DVM can be turned on, and histograms can be turned on, all the while with Wave Inspector remaining active.

When Test is pressed, the horizontal Test menu opens across the bottom of the display. The first selection is Application, and Act on Event is the default. (Previously the event was chosen). Pressing Actions, a vertical menu appears on the right, and the user can select the action that is to take place. The choices are Stop Acquisition, Save Waveform to File, Save Screen Image, Print, AUX OUT pulse, Remote Interface SRQ, Email notification, and on-screen notification. When Email notification is pressed, a menu selection Configure Email appears.

Digital Debugging and Troubleshooting

Digital devices communicate with one another and interface with digital electronic equipment by means of signals conveyed by digital buses. A defining part of the bus, though by no means the whole thing, is the medium through which the electrical pulses travel. This can consist of a pair of twisted wires as commonly seen in differential signaling, concentric cable such as coax or triax, or traces on a printed circuit board. If, with the help of a schematic diagram and/or data sheets with pinouts, you can identify the buses, you will know where to place oscilloscope probes. (The media can be checked for invisible breaks or shorts by means of quick multimeter and logic probe readings.)

In a perfect world, all digital signals would have instantaneous transitions and distinct high and low logic states. But as frequency and bandwidth increase, our measuring instruments are challenged as are our own knowledge and expertise. A good strategy is to examine, wherever possible, a good working unit that is not in failure mode. It takes almost no investment in time to create an oscilloscope acquisition, analog or digital with test point locations indicated on a schematic, save it in a flash drive (all modern digital oscilloscopes have USB slots), and save the screenshot in your computer. That way, especially if you work a lot on one type of equipment, you can map out the electronics and use this documentation when confronted with a failed specimen.

When an observed digital signal is seen to have slow transitions and/or to exhibit values outside of the ideal high and low logic states, we know that signal integrity has been impacted. The debugging process consists of detecting these problems, discovering the causes and correcting them.

Timing Problems

Signal integrity is compromised by timing errors and hardware deficiencies, either in the original design or acquired. One source of timing errors is bus contention. If this issue is not resolved, the bus may exhibit a non-threshold voltage.

Another source of timing errors consists of setup and hold violations, which are applicable in synchronous signaling. Data must be stable for a specified time interval before the clock pulse arrives. When this amount of time is insufficient, there is a setup violation. Also, after the clock pulse, the input data must remain valid for a specified time interval. If the amount of time is insufficient, there is a hold violation. If either of these violations occurs, the output may be uncertain and subject to glitches of runts. Here are some important definitions applicable in digital signaling:

Glitch has been defined as a process that has gone wrong, and you don't know the cause.

Runt is a pulse that exceeds one specified level but does not exceed a second specified level.

Metastability is an uncertain logic level due to one or more timing violations including among others nonconforming setup and hold intervals, which can adversely affect the signal output.

Notice that runt can be a type of glitch when its cause is uncertain. Either of them can be a cause of metastability.

Hardware Problems and Debugging Strategies

Design defects or post-assembly faults manifest as a variety of symptoms including fluctuations in amplitude, indistinct or misplaced rising and falling edges, data reflections, crosstalk, and ground drift.

Amplitude is affected by utility or local power supply instability, ringing, and droop and runts. Unintended capacitance or inductance can be the culprit. Irregular edges can sometimes be traced to improper board layout. Heat sources such as resistors subject to relatively heavy current have to dissipate heat. How they are located is critical. Thermal distribution, heat sinking, and active cooling should be considered. Reflections, collisions, and data loss occur when there are characteristic impedance mismatches. These situations can be corrected by changing bus terminations.

Crosstalk results when traces run parallel and close to one another for some appreciable distance. Capacitive and inductive coupling need to be quantified and reduced as needed. Fast edges equate to higher current and increased electromagnetic radiation, so these circuits may require special treatment in terms of board layout or shielding.

If a circuit's ground reference level shifts due to changes in overall current demand, voltage levels at the supply rails will fluctuate, and this can influence perceived logic levels. The remedy is to improve (reduce) ground impedance by specifying larger or shorter conductors and if needed more capable terminations.

Some Basics

When facing a digital debugging or troubleshooting job, first go into an information-gathering mode. Usually there is an articulate informant who can provide background and perspective. What is the component, circuit, or equipment supposed to do and how is it failing? Did it work properly at one time, or did the problem exist right from the start? Is it intermittent, showing up after the unit has warmed up?

The answers to these and similar questions will establish a context for an actual examination of the unit. If it is a repair job, begin by making a thorough visual inspection. Look for visible cracks in the circuit board(s) that could result in breaks or harmful impedances in the traces. Also look for unintended solder bridges between adjacent traces. In today's downsized components and more densely packed circuit boards, these faults are sometimes not visible, so additional measurements may have to be done using a high-impedance multimeter in ohms mode.

But first observe the following:

Contemporary electronics makes abundant use of CMOS technology. These circuits are based on pairs of MOSFET's, one N-type and one P-type. That arrangement is wonderfully energy-efficient since current is drawn only during very brief transitions. However, these fast rise and fall times are essentially high-frequency waveform segments that emit brief but powerful EMF spikes to the possible detriment of neighboring circuits. Another downside is that CMOS semiconductors are instantly destroyed with no visual indication if subjected to imperceptible static charge while being handled. So there is the potential for introducing additional defects that make diagnosis far more difficult. To prevent this, circuit boards should be contacted only at the edges, far away from traces and terminals. Electronic technicians use grounding bracelets to prevent static charge. Tools such as soldering irons should be solidly grounded.

Ohmmeters normally exhibit a three-volt potential between probes, which is necessary to measure resistance. This voltage should be verified on all ranges, using a second multimeter. Normally, it is safe to apply this voltage to most terminals of most components, but if in doubt, consult the manufacturer's data sheets, available online.

Before doing ohm measurements, power down the equipment and discharge (using a low-ohm power resistor, not a screwdriver) any devices that may hold a charge. On large equipment, beware of distributed capacitance.

While the equipment is powered down, you may want to pull apart and reconnect any solderless terminals or ribbon connectors. This polishes the contacts and sometimes restores operation. Some technicians touch each solder joint with a hot iron, using heat sinks as appropriate to protect components from excess heat. This also sometimes restores operation.

Then power up the equipment and, taking care that there are no hazardous voltages present, measure the temperature (without touching terminals) of all accessible components. If a semiconductor is hot, it is probably defective. If it is operating, it should be slightly warm. If it exhibits no temperature rise at all, it is either defective, receiving no power, or temporarily switched out of the circuit.

These and similar actions sometimes restore operation. To go deeper, it is necessary to acquire block diagrams, schematics, and other documentation. Carefully study power and signal flow at each component. Then, bring out the oscilloscope.

Having located the beginning and end of each digital bus, connect it to first an analog channel to check the purity of digital pulses in regard to their ability to create valid logic high and logic low levels. Then connect these buses (first input and then output) to the built-in protocol analyzer and check for glitches and runts as discussed earlier. If the equipment is operating some of the time with intermittent faults, good plan for a start is to connect the power supply to an analog channel and see if the quirky behavior can be correlated to a power anomaly.

Reference

Horowitz, Paul and Hill, Winfield, *The Art of Electronics*, Third Edition, Cambridge University Press, 2015

Chapter 7
Oscilloscope Troubleshooting Techniques

Abstract Oscilloscope troubleshooting techniques. Injecting synthesized signals at various points within stages, circuits, and devices and looking at the downstream waveforms has long been effective in troubleshooting electrical equipment and evaluating new designs. MDO and MSO oscilloscopes have far greater capabilities when time-domain waveforms are juxtaposed with the frequency-domain versions. And not to be neglected, power-quality evaluation can be performed using this oscilloscope configuration.

Electronic equipment is becoming ever more diverse and sophisticated as it penetrates into our lives and into our societies worldwide. Components are more reliable, but also more numerous and densely packed, so there is the potential for electrical faults, which can bring down a circuit, device, or entire network. It is the task of the technician to get things up and working, and it is the task of the engineer to create new designs that are less prone to faults and to make changes in existing designs to increase reliability and performance.

Fault mitigation falls into two broad categories, debugging, and repair. These activities are very similar and overlap regarding methods and procedures. Repair is applicable to equipment that has been in service but has experienced premature failure. Debugging is applicable to new prototypes that operate less than perfectly or not at all due to some design or manufacturing flaw that must be found and corrected.

Diagnostic procedures are similar but may go off in different directions depending on specifics – the nature of the equipment and of the fault that has befallen it.

A valuable approach, beginning in an information-gathering mode, is to interview the person(s) involved in operating the equipment or, in a debugging scenario, those who designed and built the prototype. We want to find out if the defect existed right from the start, or did it appear gradually or at some time (seconds, hours, days, months?) after initial power up? Is the fault total failure of all systems, or is it partial, less than perfect performance like sound distortion or noise in an audio system or intermittent performance in a digital system?

Similar lines of inquiry will suggest themselves and often go a long way toward resolving the problem. If the fault appears only after some uniform time interval

D. Herres, *Oscilloscopes: A Manual for Students, Engineers, and Scientists*,
https://doi.org/10.1007/978-3-030-53885-9_7

after power-up, it is likely heat-related, and you are well on the road to finding the solution. Sometimes the heat rise is normal, but shortly after power up expansion will cause a minute fissure in a coil to widen and interrupt the circuit.

The next stage, still in the information-gathering mode, is to assemble manufacturer's documentation, including schematics, block diagrams, parts lists, and written troubleshooting procedures if they exist. As you proceed, you may want data sheets (available on the Internet) for individual components. Especially in debugging a prototype, you may want to recommend substituting a higher-functioning semiconductor or higher-ampacity conductor (Horowitz and Hill 2015).

Getting started, you need to understand down to the component level how the equipment works and what it is intended to accomplish. Sometimes this is not necessary. A careful visual and tactile examination will suffice. A burnt component may be found. Still, the question arises, especially in debugging and prototype, why did this component fail? If it was faulty prior to assembly, OK, but if it was electrically or thermally stressed in operation, you may have a more extensive problem.

Of course many components, notably ICs and discrete semiconductors, fail without visual manifestation. Here we begin to bring out the test equipment. A high-end multimeter with a temperature probe is a valuable test instrument as is a thermal imager. When an IC fails, it is often due to an internal short. This will manifest as a short-duration burst of heat. Then the fault burns through and the short, between two pins where it shouldn't exist, becomes an open. Using the temperature probe, check all IC's and semiconductors, with equipment power on. If one of them is abnormally hot, it has an internal short (or excessive voltage or current is being applied to it). If it is excessively cold, the short has cleared and the power path is open. In any event, the component needs to be replaced. However, that may not be the root of the problem. You may have to look elsewhere to find out why excess electrical energy or none at all is being applied to the component. Alternately, especially in debugging a new design, there may be an on-board or equipment-wide heat management problem, and this has to be evaluated and resolved.

Equipment Required

The technician or engineer, for successful debugging and repair, must have a wide range of test equipment. The most versatile and always essential instrument is the digital storage oscilloscope. Don't bother with an old analog CRT scope. It is useless except for educational purposes. A modern digital oscilloscope is much more compact, lighter, and reliable with features that far surpass older models.

As for bandwidth and number of channels, this depends on what you plan to do. A brand-new, two-channel, 50-MHz Rigol digital oscilloscope can be bought for just over $300, and despite the channel count and limited bandwidth, the instrument has lots of advanced features including FFT, multiple trigger modes, pass-fail, and an excellent built-in help system. Cabinet construction is excellent.

On the other hand, four-channel, GHz bandwidth instruments with endless features are available from Tektronix, LeCroy, and Keysight (formerly Agilent). These instruments are more costly but well worth the price for advanced shops and labs and in educational settings.

Bench-type oscilloscope owners should have as a second instrument, a hand-held, battery-powered oscilloscope with inputs isolated from one another and from ground. This instrument is good for measuring voltages that are referenced to and float above ground potential, and additionally it is well-suited for field work where there is no AC power close by and for work on the factory floor where environmental conditions may be harsh.

Besides the oscilloscope, an array of test equipment is essential for successful debugging and repair. You may have a wonderful oscilloscope on the bench, but you'll pick up a multimeter for quick, highly accurate measurements on a dense printed circuit board. These two instruments are often used in conjunction with one another, sometimes for simultaneous readings. There is often use for more modest tools such as the neon test light, incandescent appliance bulb with plug-in socket, logic probe, and pulser.

At the other end of the price continuum, actually exceeding the cost of the oscilloscope is the spectrum analyzer, described earlier. This instrument is optimized to image spectral distribution in the frequency domain, and it incorporates highly specialized analytic tools that far exceed the FFT and frequency domain capabilities found in the best oscilloscopes.

Other instruments, essential in digital work, include:

- Network Analyzer – Measures network parameters
- Signal Analyzer – Measures amplitude and modulation of an RF signal
- Bus Analyzer – Different models are designed for specific buses
- Logic Analyzer – Captures and displays multiple buses
- Protocol Analyzer – Captures and analyzes signals and data over a communication channel

Some features of these instruments are available to various extents in a digital storage oscilloscope.

The oscilloscope is a universal, highly capable instrument and plays a central role (alongside the multimeter, which is also indispensible) in electronic equipment debugging and repair.

Signal Generator: AFG vs. AWG

In debugging new designs and diagnosing existing electronic equipment, the oscilloscope is frequently used in conjunction with a signal generator. In analog diagnosis the basic idea is to inject an appropriate signal at or near the upstream (signal input) end and to connect oscilloscope probes at various points downstream so that the modified waveform can be observed. There are variations on this generic procedure, as when we check the output of an oscillator without the signal generator connected.

The success of this procedure depends among other things upon the correct placement and use of the signal generator. With the introduction of digital instrumentation, signal generators gained greater functionality and, along with digital oscilloscopes, greatly facilitated debugging and system diagnosis.

Modern signal generators come in two varieties, arbitrary function generators (AFG's) and arbitrary waveform generators (AWG's). They have in common the ability to capture or synthesize user-designed waveforms of varying complexity. In addition, both instruments contain libraries of common waveforms such as sine, square, pulse, ramp, triangle, DC, noise, sin(x)/x, Gaussian, Lorentz, exponential rise, exponential decay, Haversine, and cardiac. Numerous parameters can be set including frequency/period, amplitude, and offset.

Arbitrary is the key word in these instruments. By means of menus and associated controls, the user can begin with any of a number of idealized waveforms and insert points to alter the trace to create a vast number of customized waveforms, which can be stored in memory and later used to test equipment under varying input and environmental conditions.

The AFG is the more used of the two types, in part because it is the less expensive alternative. It has a somewhat more limited waveform library, but this instrument has fast frequency response and excellent stability. For most work, the sine and square waves plus several others are sufficient, and the ability to quickly change frequencies is a plus. Most modern digital storage oscilloscopes include built-in AFGs, which are very convenient (as opposed to having a separate signal generator on the bench). The features are less extensive than those of an autonomous instrument, but for many applications they are well-suited. Waveforms can be exported beyond the host machine by means of a BNC cable from AFG Out in one instrument to an analog or RF port in another instrument.

AWG's surpass AFG's in the ability to create almost any waveform that can be imagined. Memory segmentation and waveform sequencing are useful in specialized applications, but for most debugging and diagnosis, in conjunction with an oscilloscope, the AFG is the usual choice.

Digital Analysis

Mixed-domain oscilloscopes such as the Tektronix MDO3000 series are equipped with built-in MSO-capable 16-channel logic analyzers, so in this respect they are in the same league with the more advanced MSO5000 series instruments. All of these oscilloscopes, in the digital mode, view a digital signal as either logic high or logic low, in the same manner as a digital circuit views the signal. As long as distortion, noise, ringing, overshoot, or ground bounce are not great enough to disturb transitions, these aberrations are invisible to the oscilloscope in the digital mode. To see and evaluate them, the signal must be applied to an analog channel. In the digital mode, the instrument uses a threshold voltage to determine if the signal is logic high or logic low. These threshold settings are useful in debugging circuits with mixed logic families.

Signal Tracing a Radio or TV

An oscilloscope used in conjunction with a signal generator is a very efficient method for locating an electrical fault. For simplicity, we'll consider an FM receiver, but these methods are applicable for many types of electronic equipment other than receivers.

To begin, you'll want product documentation with a schematic, block diagram, and parts list with components values and parts numbers. In the real world, this information is not always available. In that event, the procedures are the same, but there is the added task of analyzing wiring, ribbon connectors, and circuit board traces in order to identify the stages. The placement of stages and components generally does not conform to signal flow beginning at the antenna and proceeding to the speaker because the actual products are designed for optimum cooling, resistance to electromagnetic interference, characteristic impedance, and other requirements.

In this discussion, we'll assume a preliminary assessment has been completed. A visual inspection has identified blown electrolytic capacitors, burnt resistors, cracked circuit boards, and the like. Good DC is coming out of the rectifier. You can test a speaker by connecting a C-cell to see if the cone moves. Soon, you've concluded that the fault is likely to be a non-visible component or solder joint failure.

In troubleshooting, a priority should be to avoid introducing an additional fault, which would make the final repair orders of magnitude more difficult. Modern electronics is making increased use of CMOS technology, which is based on complementary pairs of N-type and P-type semiconductors having insulated gates. Rather than electrical current, the signal input consists of a fluctuating electrical field that controls the output. This arrangement is highly efficient from the point of view of power consumption and immunity to noise, but the problem is that the very thin insulating barrier is instantly destroyed (leaving no visual indication) if a slight electrostatic charge is applied to the terminals. Such a charge is imparted by the unknowing technician's touch, and an additional fault is introduced.

Another problem is that if a powerful signal is introduced, an expensive speaker may be blown. The solution is to temporarily disconnect the speaker, replacing it with a dummy load consisting of a high-watt power resistor, matching the resistance (often 8 ohm) stamped on the speaker. Moreover, this substitution should be made with the set powered down, because operating the amplifier with no load can cause the voltage to go high in the final stage output circuit, destroying the transistor(s).

In commencing the signal-tracing process, the oscilloscope will not be able to see a signal at the antenna. The amplitude is below the range of ordinary measuring instruments. For the oscilloscope, the first signal will appear at the tuner output. The tuner is usually in a separate grounded metal cage whose purpose is to exclude electromagnetic interference. Don't expect to make any repairs to the tuner. Due to the small amplitude of signal in this area, there is rarely a fault here. Moreover, because of their sensitive nature, tuners are generally sent out to a specialized shop for servicing.

Signal tracing can be done using a broadcast signal from the tuner as input. But the better method is to rely on the signal generator. There are two reasons. For one thing, the tone is more uniform, and its frequency and degree of modulation can be controlled and monitored more accurately by the oscilloscope at various points. And secondly, the tone can be injected at different locations.

To get started, you may want to check the output of the local oscillator. This is done by looking at that signal with the signal generator injecting a tone at the tuner. The purpose of the oscillator signal is to combine with the RF to make a lower, more manageable intermediate frequency (IF) carrier, still modulated. The oscillator frequency varies as different RF frequencies are tuned in (remember the old double-gang variable capacitor) so that the IF remains the same for all broadcasts, which is one of the rationales for its use.

After verifying oscillator operation, you can begin signal tracing. You can start at the first stage after the tuner. Leave the signal generator in place and move the oscilloscope downstream stage by stage, until the signal disappears. You've found the defective stage!

To speed things up, you may want to start the oscilloscope somewhere in the middle and, using the half-splitting technique, find the bad stage in fewer steps.

Once the bad stage has been found, you can work inside that stage to isolate the bad component. In fact, the signal generator-oscilloscope combination can be used to check a single transistor or IC, once you've identified all those pins from the data sheets.

The signal generator and oscilloscope can be connected in numerous configurations. For example, two or more stages can be checked together. In a stereo set with one bad channel, comparisons can be made. Similarly, if you have a known good working set, readings can be compared.

The foregoing is an introduction. Besides having a dead stage as described, there may be distortion issues, or intermittents. Television is far more complex, because in addition to audio, there are video signals, synch, color bursts, and other specialized circuitry that differs depending on the make. One useful feature of the schematics is that they incorporate, at strategic locations, small graphics showing what an oscilloscope display should look like in a good working set.

Debugging Digital Buses

Gone are the days of the easy to understand but inefficient parallel bus (16 wires, 32 terminations). Now we have the efficient serial bus (two wires, four terminations) that is far more difficult to learn. First you have to understand the fundamentals and then figure out how to debug it. On top of that, there are at least a dozen primary serial buses with more coming down the road, each with its own architecture and priorities.

Of the numerous conductors in a parallel bus, each signal component has its own path. These separate lines may include as many as 16 address lines, 16 data lines, a clock line, and additional control signals. Address and data values are conveyed over the parallel lines simultaneously. State or pattern triggering present in most oscilloscopes makes it a simple matter to trigger on the event of interest. It is not difficult to understand the captured data.

Unfortunately all these wires created a massive hardware burden. Bus designers saw the need for serial communication, where the same information is conveyed over fewer lines, typically two and sometimes one. In this new serial technology, a single signal may contain address, data, and even clock information.

As an example, a CAN bus serial signal contains a start of frame, identifier or address, data length code, data, cyclic redundancy check (CRC), and end of frame, and other information. The clock is embedded in the data, and bit stuffing ensures sufficient edges for the receiver to lock onto the clock. The hardware reduction has been beneficial, but designer and debugger, now in the serial regime, require far greater sophistication to work within this new reality. You can't easily interpret the message just by looking at it. If such a message is faulty and occurs only once every few hours, there is a serious problem that presented great difficulty for traditional oscilloscopes.

The same problem arises even with the far simpler I²C serial bus. Since I²C uses separate clock and data lines, the clock is available as a reference. But it is still necessary to ascertain where data goes low while the clock is high in order to find the start of a message. Then you have to manually inspect and record the data values on clock rising edges and organize the bits to structure the message. And this process has to be repeated a great many times.

New oscilloscope technology involves optional serial triggering and analysis capabilities. To access them, in the front panel controls press B1 or B2 (depending on how you are probing the equipment under investigation). In the bus menu that appears across the bottom, press Bus. (Parallel is default.) Serial buses that are available are I²C, SPI, RS232, CAN, LIN, FlexRay, Audio, USB AND MIL-STD-1553. Any of them can be selected by using Multipurpose Knob a.

I²C

In I²C, for example, the horizontal menu that appears across the bottom allows these choices:

- Define inputs – four analog channels and 16 digital inputs.
- Thresholds – SCLK voltage set by Multipurpose Knob a and SDA voltage set by Multipurpose Knob b. Additionally, TTL, CMOS, and other logic family voltages can be set automatically by pressing the soft key associated with User Preset.

- Other menu selections permit the user to label the signals, choose bus display, choose bus and waveforms, choose Hex or choose Binary, and turn on the Event Table.

Other serial bus types display different menu choices as appropriate.

When you have defined the clock and data channels plus logic 0 and logic 1 thresholds, the oscilloscope can interpret the material that is conveyed through the bus. It is now possible for the instrument to trigger on message-level information and decode the acquisition in an accessible manner.

SPI

We can define an SPI Bus by entering the bus parameters including which channels SCLK, SS, MOSI, and MISO are on, in addition to thresholds and polarities. This is similar to the I²C process, but some of the details are different.

Consider an embedded system. An SPI bus is connected to a synthesizer, a DAC, and I/O. The synthesizer uses a VCO that also provides a system-wide 2.5 GHz clock signal. The synthesizer is normally programmed by the CPU at startup. But there is a problem. The VCO is stuck at its rail generating 3 GHz. The best approach is to look at the signals between the CPU and the synthesizer. These signals appear to be present.

Then we check the information going across the SPI bus to program the synthesizer. The procedure is to configure the oscilloscope to trigger on the synthesizer's Slave Select signal and power up the embedded system to display startup programming commands. To determine if the device is programmed correctly, we consult the data sheet for the synthesizer. According to this document, the last hex character in the first three transfers should be 3, 0, and 1, as opposed to 0, 0, and 0.

The error then, is in these final three digits in each 24-bit word. It is a simple matter to make the correction in the software.

USB

Modern digital oscilloscopes can perform USB serial triggering and analysis. USB trigger, decode, and search support is provided, but for high-speed USB, the oscilloscope bandwidth has to be at least 1 GHz.

To check data latency performance of a full-speed memory device, the procedure is to check to determine if the memory device responds to the computer IN token request for data. To do this, a differential probe is connected between the computer and USB memory device.

First go to the bus menu in the oscilloscope, and choose USB from the list of supported standards to define the USB Bus. Then, to define the parameters, select menu choices: speed, source channels, probe type, and thresholds.

Check the enumeration process by triggering on the Setup token. Then, verify Start of Frame packets. Trigger on them and check the speed by looking at the J idle state to see if it is positive.

Next, set up the oscilloscope to trigger on a NAK token and shift the oscilloscope to Single Acquisition. This accesses the computer request data from the memory device. The oscilloscope will not trigger if the memory device is ready to transfer data. Otherwise, it will send a NAK in response to the computer host IN token, and the oscilloscope will trigger on the NAK.

The user can copy oscilloscope trigger settings, which then become search criteria for Wave Inspector. It will then search the acquisition looking for all NAK's. These NAK's can be viewed by pressing Wave Inspector Next and Previous buttons.

Ethernet

Because one differential Ethernet signal contains address, control, data, and clock information, identifying pertinent events can be difficult. However, new-generation oscilloscopes are capable of debugging 10-BASE-T and 100-BASE-TX systems using automatic trigger, decode, and search.

The procedure begins by having the oscilloscope trigger on Ethernet packet content such as Start Frame Delimiter, MAC addresses, MAC length and type, MAC client data, Q-tag control information, IP header, TCP header, TCP/IP4 client data, End of Packet, Idle, and FCS (CRC) errors.

A decoded display appears and it provides a view of individual signals. The user can see where packets begin and end and can identify sub-packet components. Each packet is decoded and the values are displayed in familiar formats.

Packets are time-stamped and listed in columns. This serial triggering isolates the event of interest. The next step is to use Wave Inspector to search through the data for user-defined criteria, especially serial packet content. Wave Inspector highlights each occurrence. Press Previous and Next to navigate.

RS-232 Bus

As in other serial buses, triggering and analysis for RS-232 is available in modern digital oscilloscopes. In the usual manner, define RS-232 by entering the parameters, in this instance channels to be used, bit rate, and parity. RS-232 displays data as binary, hex, or ASCII.

If a device that polls a sensor for data over an RS-232 Bus is not functioning, that is the sensor is not responding to requests for data, possible difficulties are that the sensor is not receiving the requests or that it is receiving the requests but ignoring them. The procedure to test for this condition is to probe the Tx and Rx lines and set up the bus on the oscilloscope. Set the oscilloscope to trigger on the request.

Zooming in on the waveforms to see the response from the sensor, it may be that the controller's timeout is insufficient for the sensor to reply, so increasing the time-out restores the RS-232 Bus functionality.

CAN Bus

To enable CAN Bus serial triggering and analysis, press the oscilloscope front panel Bus buttons to define the bus. The parameters are type of CAN signal, input channel, bit rate, threshold, and sample point as a percent of bit time. The desired frame can be triggered by specifying the ID of the CAN module and associated command data.

Markers as reference points can be placed on the waveform. They are added or removed by pressing the Set and Clear Marker buttons on the oscilloscope front panel. Pressing Previous and Next causes the zoom window to move to adjacent markers, revealing events of interest in the acquisition.

LIN Bus

LIN Bus is a one-wire implementation, so serial triggering and analysis in an oscilloscope is a fairly simple matter. The bus is defined by entering parameters such as the LIN version that is used, the bit rate, polarity, threshold, and location of data that is sampled.

LIN triggering capability in the oscilloscope includes triggering on the sync field, triggering on a specific identifier, triggering on data values or ranges, triggering on identifier and data, triggering on a wakeup frame, triggering on a sleep frame, and triggering on sync errors, ID polarity errors, and checksum errors.

Elements of interest can be isolated on a LIN Bus using the above triggering modes.

MIL-STD-1553

Serial triggering and analysis for MIL-STD-1553 can be performed in a modern digital oscilloscope by using passive probes on the analog channels. Parameters are entered in horizontal menu options as in other bus types. These include Sync, Word Type, Data Word Value, and Parity Error.

FlexRay

FlexRay is a differential bus. Media is shielded or unshielded twisted pair. Serial triggering and analysis is defined in the bus menu by selecting FlexRay.

Then, in the Define Inputs menu, the user chooses Channel A or B and the signal type. Next, set the thresholds and bit rate.

Triggering capability for FlexRay includes start of frame, indicator bits, identifier, cycle count, header fields, data, end of frame, and error.

Audio Buses

To define audio bus, enter the basic parameters including word size, signal polarities, bit order, and thresholds. Then, the user triggers on specific bus data content, decodes acquisitions, and searches through them to find the desired data.

Audio triggering capability includes word select, frame sync, and TDM data. These trigger criteria are also available as search criteria for investigating long acquisitions. The decoded audio data can be presented in event table format.

Digital Debugging Notes

Some electronics technicians and engineers work exclusively in the analog realm. Others prefer digital circuitry. Each subgroup holds that their chosen field is easier to understand and that it conforms to a simpler and more uniform set of principles.

In reality, to debug new prototypes or to diagnose equipment that has failed after performing as expected for some period of time, knowledge, and expertise in both fields is required.

Digital circuits should be probed and observed in the analog domain as a prelude to digital analysis. That way, if the digital pulses are corrupted to the extent where they cannot reliably convey logic 1 and logic 0 voltage levels, that problem perhaps due to an erratic power supply, will have to be corrected before it is worthwhile to proceed.

To get started in the digital mode, these are the tasks:

- Probe the digital output and locate any anomalies.
- Isolate specific events that seem out of place or may be causing faulty operation.
- Use cursors and automated measurements to quantify digital output.
  To summarize, most digital circuit problems can be readily debugged. The process includes these steps:
- Ascertain the fault by observing the digital display. It may consist of infrequent events such as runts, glitches, or other transient events.

- Place the anomaly in the instrument's memory. Using the oscilloscope's flexible triggering capability, you can display the anomaly and discover its cause. Use Record Length to determine the time interval that is to be captured. (This is set by pressing Acquire, then the soft key associated with Record Length, which is adjusted by turning Multipurpose Knob a.)
- In Wave Inspector, press Search to find the event. The horizontal menu at the bottom of the display permits the user to choose a search type, which may be Edge, Pulse Width, Timeout, Runt, Logic, Setup, and Hold, Rise/Fall Time and Bus. Also, Source, Slope, and Threshold may be set.
- To find the cause of the observed anomaly, use waveform analysis tools including Automated Measurements, Statistics (accessed by pressing Measure in the Wave Inspector section), and the Cursor button at the top of the front panel.

To capture a digital signal, it is necessary to set the threshold correctly. This voltage level determines the point above which a digital signal denotes logic high and below which it denotes logic low. This process functions reliably for large signals that transition rapidly between the logic high and logic low levels. If the voltage lingers at all at the threshold level, the result is an indeterminate digital state and unpredictable circuit behavior.

The answer is to use two distinct voltage levels in order to create a hysteresis situation whereby the rising and falling transitions are separate by a reliable time interval. In a clocked circuit, the transitions must not occur during a minimum setup time before the clock edge and a minimum hold time after the clock edge. When these intervals are not maintained, circuit behavior will be impacted. The conditions are known as Setup and Hold Violations.

The oscilloscope can easily be configured to search for Setup and Hold violations in Wave Inspector and to measure and analyze them with a view to mitigating the underlying cause(s) and restoring correct operation.

Pulse Stream

The above measures depend upon capturing the pulse stream, and that is what we will consider now. To demonstrate, we connect a Tektronix MDO 3000 Series of equivalent oscilloscope to the Tektronix Demo One Board. (I have found that oscilloscopes can be connected to demo boards made by competing manufacturers with no ill effects as long as voltage limitations and floating grounds are observed.) Be sure to use the specialized dual USB cable that is supplied. That is because the Demo One board draws an amount of current that could damage a single oscilloscope USB slot and circuitry. The dual Type A USB connectors must be plugged into two separate USB slots in the oscilloscope.

Connect a 10:1 probe to the Channel One analog input. To begin, we'll probe the frequent anomaly digital signal in the Demo One board. Attach the ground return lead to the ground terminal on the board and the probe tip to the frequent anomaly terminal. Power up the oscilloscope. The board also has a power button. The board's green power LED should be on, indicating that the board is powered up.

On the oscilloscope, set the horizontal scale to 200 ns/div. In the oscilloscope triggering section on the front panel, set the trigger level to 2 Volts. Press the front panel trigger button to Single. Press Menu Off to reveal the entire display.

Alternately press Run and Single to discover if anomalies are present in the signal. These will consist of abnormally high or low values and abnormally narrow pulses. Press Run/Stop to create a dynamic display. You now see what looks like a single waveform, but it actually consists of hundreds or thousands of superimposed waveforms. Included among them are relatively faint (because they are less frequent) narrow glitches and reduced amplitude runts.

In debugging, it is important to keep in mind the definitions of glitches and runts, since these are what are being sought. A glitch is an anomaly, typically a narrow pulse, that as yet has no known cause. A runt is a signal whose amplitude rises above a user-specified lower level but fails to rise above a second user-specified higher level.

Since you have now detected and observed representative anomalies, you can proceed to the next step, which is to capture them using specialized digital debug tools in the oscilloscope such as pulse width glitch triggers and runt amplitude triggers.

To proceed, we need to capture an observed narrow pulse. Since we previously set the horizontal scale at 200 ns/div and the narrow pulses appear to be a little less than one-quarter division wide, we'll set pulse width triggers to capture all pulses that are narrower than 50 ns. To do this, press the Trigger Menu button. In the horizontal menu that appears across the bottom, press the soft key that is associated with Type. Then, in the vertical menu, use Multipurpose Knob a to select Pulse Width. Press the soft key associated with Trigger When and set the pulse width value to less that 50 ns. Press Menu Off to remove the side menu. Again, press the Single button.

With the positive pulse width trigger, you can easily capture each of the narrow positive pulses in the signal, shown in the center of the display next to the orange T icon. Noted that with pulse width triggering, a single 2-Volt threshold value is used, and the pulse width is measured at that voltage.

We shall now proceed to runt pulses. Press the soft key associated with Type. Use Multipurpose Knob a to select Runt as the trigger type. Then, press the soft key associated with thresholds. Set the high threshold at 2 Volts and the low threshold at 1 Volt. This defines the runt we want to capture. Press Menu Off and the Run/Stop button.

With the runt trigger, you can easily capture each of the positive runt pulses in the signal, shown in the center of the display next to the orange trigger T icon.

Now that we have isolated a single glitch and a single run, we'll move on to Wave Inspector in order to reveal all such instances. First, using the horizontal scale control, set the horizontal scale to 100 µs/div. Then press the Single button at the top of the front panel. In the display, hundreds of waveforms are packed tightly together, so details cannot be discerned.

Turn the inner Wave Inspector knob clockwise to activate zoom. Adjust the degree of zoom to clearly display the runt pulse at the trigger point. Turn the outer

Wave Inspector knob to move the zoom window from side to side. As you turn the control farther from center, the panning speed increases.

At the top of the screen, we see the entire waveform at the bottom, the zoomed region is shown in brackets. Press the Play/Pause button to pan through the acquisition. To find another runt, use the outer Wave Inspector knob to change the panning speed and direction. Press the Search button in the Wave Inspector section to find and mark characteristics of interest on the waveform. The markers are white triangles at the top of the display.

Press the soft key associated with Search in the horizontal menu at the right. Turn Search on. In the vertical menu, press the soft key associated with Copy Trigger Settings to Search. This copies your runt trigger setup into the search engine. White triangles appear at the top of the display. These are runt events in the acquisition.

You can toggle back and forth between these events by pressing the front panel < and > arrow buttons below the concentric Wave Inspector knobs. Be sure to turn off Search when you are done so it doesn't interfere with subsequent operations.

Finally, you will want to analyze digital signals using manual and automated measurements. To do this, in Wave Inspector, turn off zoom and remove all menus. Set the horizontal scale to 200 ns/div. Press the Run/Stop button. Press the front panel Cursors button to turn on vertical bar cursors.

Cursors perform manual measurements, such as quantifying the width of a pulse. Notice that the cursor measurements are shown in the cursor readout in the upper right corner of the display. Similarly, amplitude can be measured.

As can be seen, cursors are an effective way for the user to manually measure wave parameters. However, automated measurements are another way to go. They are quick and highly accurate.

To perform automated measurements, press Measure in the Wave Inspector section. Press the soft key associated with Add Measurement in the bottom horizontal menu. In the vertical menu, press OK Add Measurement. Use Multipurpose Knob a to choose positive width. Press Add measurement. Notice that a maximum of four measurements can be retained. To add more measurements, the appropriate number of prior measurements must be removed from the list.

To focus measurements on a specific portion of the waveform, such as a runt pulse, use the cursors to gate the measurements. In the bottom horizontal menu, under More, press More repeatedly until Gating is selected. Then, in the vertical menu to the right, press the soft key associated with Between Cursors.

Using Multipurpose Knob a, move the a cursor slightly to the left of the runt pulse. Using Multipurpose Knob b, move the b cursor slightly to the right of the runt pulse. Notice that the measurement now reflects the width of the runt pulse.

Analog Diagnostics

An increasing amount of electronic equipment currently combines digital and analog circuitry. A case in point is the familiar three-phase induction motor configured with a variable frequency drive.

Debugging and diagnosing analog circuits is probably neither more difficult nor less difficult than doing digital work. It just requires a different mindset and expertise. In both realms, the oscilloscope figures prominently. In the discussion that follows, we'll once again bring out the versatile oscilloscope, using it this time to focus on analog debugging and diagnostic procedures.

Reference

Horowitz, Paul and Hill, Winfield, The Art of Electronics, Third Edition, Cambridge University Press, 2015

Chapter 8
Oscilloscope Networking and Device Communications

Abstract Oscilloscope networking and device communications can be accomplished using a variety of techniques including USB and Ethernet, LAN, E*SCOPE, Visa Drivers, LABVIEW, MATLAB and LXI. Pinging and External Video Feed are discussed.

This chapter focuses on oscilloscope networking and similar functions. We'll use this term broadly – not just connecting into a computer network, but all the many ways information can be extracted from an instrument of origin and conveyed elsewhere or the ways an oscilloscope can receive and act on commands from a remote computer or other device. There are numerous modes, beginning with the modest USB and Ethernet ports. These simple outputs permit the oscilloscope to communicate with other test and measurement instrumentation or a computer, using a flash drive or various cable media.

A quick and easy way to extract a waveform or settings from the oscilloscope is to insert into the oscilloscope USB slot a flash drive (also known as thumb drive and memory stick) and press the Save button on the oscilloscope front panel. Then, transfer the flash drive to a computer USB slot. The contents of the flash drive can then be saved into the computer memory, stored indefinitely, copied to a CD, emailed to a colleague or whatever is required.

It is worth noting that to properly remove a flash drive from a computer, it must first be ejected. To remove it from the oscilloscope, however, this step is not necessary, and in fact there is no provision for doing so.

The flash drive has many advantages over the obsolete floppy disc and even over the newer optical disc. It is small, inexpensive, reliable, and has enormous storage capacity, as much as one TB. Depending on the memory chip, up to 100,000 write/erase cycles are possible, with anticipated hardware life approaching 100 years. With no moving parts, the flash drive is highly durable, and it is not subject to electromagnetic interference.

© Springer Nature Switzerland AG 2020
D. Herres, *Oscilloscopes: A Manual for Students, Engineers, and Scientists*,
https://doi.org/10.1007/978-3-030-53885-9_8

197

Because it is USB-powered, the flash drive requires no batteries or external power supply. USB mass storage device class is used, supported by Windows, Mac OS, and Unix operating systems.

To get started, display a waveform on the oscilloscope screen, insert a flash drive and press Save. *(In some oscilloscopes, Save will be found behind several menu layers. Also, rather than Save, some manufacturers such as Rigol use the print button for this function.)* Transfer the flash drive to the computer USB slot, and, depending on the settings, a clickable flash drive icon will appear on the computer desktop, so that the waveform can be made to display on the computer screen. For full functionality, however, you need to go into the oscilloscope settings. The following is offered as a guide:

Setups, waveforms, and screen images can be saved permanently inside the oscilloscope, or they can be saved to a flash drive. To save material to a flash drive, select the appropriate menu such as the To File side menu. Use Multipurpose Knob a to scroll through the external file choices. E is a flash drive that has been inserted into the USB slot on the oscilloscope front panel. F is the flash drive inserted into the USB slot on the oscilloscope front *or rear* panel. Use Multipurpose Knob a to select the desired file.

You can name a file or use the default file name. (Of course after it is in the computer, you can always rename a file.) Here are some Tektronix file names and extensions:

Tek00000 through Tek99999.png, .bmp, or tif are image files.

.csv denotes spreadsheet files. .isf denotes internal format files.

For waveforms, numerals are 0000 through 9999, followed by three additional characters, which denote, CH 1, CH 2, CH 3, or CH 4.

MTH for a math channel RF 1 through RF 4 for reference memory waveforms ALL for a single spreadsheet file containing multiple channels when Save Waveforms is selected.

For RF traces, there are four numerals. Additionally,

NRM denotes a normal trace.
AVG denotes an average trace.
MAX denotes a maximum hold trace.
MIN denotes a minimum hold trace.
TIQ denotes a baseband I and Q file.

Editing File, Directory, Reference Waveform, or Instrument Setup

The following procedures are applicable to a Tektronix oscilloscope. Instruments by other manufacturers are similar, but the terminology differs: Press Save/Recall Menu. In the menu that appears across the bottom, press Save Screen Image, Save

Waveform, or Save Setup. Then press the soft key associated with Save Screen Image in the horizontal menu. In the vertical menu on the right, the first menu item is File Format. Pressing the associated soft key toggles through .bmp, .png, and .tif. The second soft key enables Ink Saver Mode, which has a printer-friendly white background.

The second menu item on the bottom is Save Waveform. This time, in the vertical menu on the right, the user can turn Multipurpose Knob a to select the source and Multipurpose Knob b to select the destination, reference channels R1, R2, R3, or R4, or the flash drive if it is still mounted in the USB slot.

The third menu selection is Waveform Gating, selected by turning Multipurpose Knob a. Gating pertains to the portion of the waveform that is to be saved, which can be Full Record when gating is off, Full Screen, or between cursors. After these parameters have been set, the user can press OK Save to complete the operation.

The Recall menu permits the user to display previously saved waveform and setups. To recall waveforms, press the associated soft key in the horizontal menu. In the vertical menu are dated and time-stamped reference files, accessed by pressing the soft keys. These reference files are stored in non-volatile memory, so they are retained even after the oscilloscope has been power cycled. A maximum of four waveforms can be saved. Additional waveforms overwrite existing reference files, starting with the oldest.

The next item in the horizontal Save/Recall menu is Recall Setup. Pressing the associated soft key brings up a vertical menu. Ten dated and time-stamped setup files can be accessed. The current setup file can also be saved to a flash drive or to a mounted network drive.

To toggle on and off a reference waveform in the display, press R, just below the Math button. R-1 through R-4 appear in the horizontal menu at the bottom of the screen. Waveform setups include vertical, horizontal, trigger information and Measure capability.

A waveform can be saved by pressing the Save button a single time if the parameters have been previously set. If a waveform has been directed to a flash drive, successive Save button operations can be executed without redefining the flash drive as the target.

To manage drives, directories and files, start by pressing the Save/Recall Menu button. Then press the soft key associated with File Utilities. A vertical menu permits the user to execute any of the following actions:

Create a new folder.
Delete a highlighted file or directory.
Copy a highlighted file, directory, or drive.
Paste a file, directory, or drive that has been copied.
Rename a file, directory, or drive that has been highlighted.
Connect a networked drive.
Format a drive that has been highlighted.

The user can connect a computer to the oscilloscope to save setups, waveforms, and screen shots directly to a drive and to recall waveforms and setups from the drive. This step may involve contacting the network administrator.

After a network connection has been established, press the Save/Recall Menu button on the front panel of the oscilloscope.

Connecting Oscilloscope to Computer via LAN

E*Scope (this is Tektronix terminology) permits the user to create a Local Area Network (LAN) so that oscilloscope and computer can communicate. This program has a lot more functionality than waveform and data transfer by means of a flash drive. For one thing, the computer actually controls the oscilloscope, and furthermore the successive screens are dynamic rather than static.

Using E*Scope the connectivity is web-based. The oscilloscope contains an internal server, and of course the computer has a web browser so all you need to do is establish an Internet connection. It follows that E*Scope is cross platform. If you are a Mac person, you're OK.

E*Scope software comes on a disc with the oscilloscope, or it is available free of charge on the Tektronix website. After it is installed on the computer, you are (almost) ready to go. But first you'll need a signal. The sine wave from the oscilloscope's internal AFG is suitable. Just run a BNC cable from AFG Out on the back panel to one of the analog channel inputs on the front panel. Leave the AFG off for now. Run an Ethernet patch cord from the IP modem to the oscilloscope Ethernet port on the back panel.

In the oscilloscope, press the Utility button below the screen and then the soft key associated with Utility. In the vertical menu on the left, use Multipurpose Knob a to select Input/Output. In the horizontal menu at the bottom press the soft key associated with Ethernet and LXI.

In the screen that appears is the network IP address. Write it down and keep it in a secure place for future reference. If, however, you have changed locations and have a different Internet provider, the oscilloscope will have a different IP address, so you will have to start over.

The vertical menu on the right has a menu item labeled Test Connection. If the connection has failed, it may be reset.

Turning to the computer, enter the IP address in the browser's address bar and hit Enter/Return. Turn on the AFG. In the computer, click on the control tab at the top of the E*Scope web page. The displays on the oscilloscope and computer screens correspond. At the bottom of the computer screen are some clickable tabs that correspond to controls in the oscilloscope front panel.

The interesting thing is that the oscilloscope can be controlled from the computer. For example, press Measure in the horizontal controls at the bottom of computer display. In the horizontal menu, press Add Measurement and Remove Measurement. These actions in either the oscilloscope or the computer have the same effect on both displays. Press Waveform Histograms >More (1 of 2), and in the vertical menu, turn Waveform Histograms off. The associated soft key toggles between Off, Vertical, and Horizontal. The computer manipulates the oscilloscope controls.

At the bottom of the computer display, click on Measure. Then, use Multipurpose up and down triangles to select between Frequency and Off. The large triangle navigates from the first item to the last item in the AFG menu. The medium triangle takes you to the midpoint in the list. The small triangle does single steps.

Connecting Oscilloscope to Computer Using Visa Drivers

Another way to connect an oscilloscope to your computer is by means of Visa. This does not depend upon a web-based LAN as in the E*Scope implementation, although it is possible to configure an optional Internet connection. The two instruments are directly connected through USB or Ethernet with no connection to the IP modem unless desired.

Visa is an industry-wide driver package that enables the oscilloscope-computer connection. TekVisa is a Tektronix version optimized to work with Tektronix instrumentation. It works with MS-Windows computers, so Mac users are not able to use this method for the time being.

To set up this connection, run USB or Ethernet cable between the two instruments. Then, load into the computer the Visa drivers and OpenChoice desktop software, available on disc or free of charge as a download from the Tektronix website.

Press the Utility button at the bottom of the front panel, then the soft key associated with Utility Page. In the vertical menu that appears at the left of the display, use Multipurpose Knob a to scroll down to I/O.

If you are using USB to connect the oscilloscope to the computer, the system sets itself to establish the connection, provided USB is enabled. If USB is not enabled, press USB in the lower horizontal menu, and Connect to Computer in the vertical menu at the right of the display.

If you are using an Ethernet connection rather than USB, press Ethernet and LXI in the horizontal menu at the bottom of the display. Use the soft keys associated with the vertical menu at the right to adjust parameters as needed.

If you are using Ethernet cable such as Cat 5e to connect the oscilloscope and computer, it has to be a crossover cable rather than a standard cable. You can purchase a pre-made crossover cable or, if the distance is long, say from one room to another, you can obtain the correct length of cable and put your own male connectors on the two ends, using an inexpensive Ethernet crimping tool and the instructions that come with it in order to create a crossover cable. Both ends should be marked with an X so that in the future someone does not attempt to use it as a standard Ethernet cable.

Why crossover? The answer is very simple. If you used a standard Ethernet configuration at each of the two ends, you would be connecting the two transmit pins together and the two receive pins together, which would not work. You need to connect the transmit pins to the receive pins. Standard configurations at both ends are used when you are going from a modem to a computer, or when there is an Ethernet hub at one end. The crossing over is taken care of in the hub or modem. Going from device to device such as in the oscilloscope to computer networking project, we are discussing, or when connecting two computers, the crossover configuration is needed at one end (not both ends, or you are back where you started from). The crossover configuration connects the 1,2 pair on one end to the 3,6 pair on the other end.

RJ-45 connectors are used at both ends. They are similar to the familiar RJ-11 telephone connectors, but larger and not compatible. Since the Ethernet cable has eight conductors (four pairs), not all of them are used. However, it is customary to crimp all of them into the connectors for durability.

Notice that the individual pairs are twisted. This is essential. In conjunction with differential transmission, it eliminates interference, crosstalk, and noise. Where the cables are crimped into the connectors, this twisting is eliminated. Therefore it is desirable to keep these untwisted segments as short as possible.

Rather than making a crossover cable, a stand-alone Ethernet hub can be used, in which case straight-through Ethernet cables connect each device to the hub. If at any time the connection fails, the first thing to check is that the hub is getting AC power, indicated by an LED. The simplest thing to do is to locate the devices close together and use a factory-made Ethernet cross-over patch cord.

If you are using Ethernet cable such as Cat 5e to connect the oscilloscope and computer, it has to be a crossover cable rather than a standard cable. You can purchase a pre-made crossover cable or, if the distance is long, say from one room to another, you can obtain the correct length of cable and put your own male connectors on the two ends, using an inexpensive Ethernet crimping tool and the instructions that come with it in order to create a crossover cable. Both ends should be marked with an X so that in the future, someone does not attempt to use it as a standard Ethernet cable.

RJ-45 connectors are used at both ends. They are similar to the familiar RJ-12 telephone connectors, but larger and not compatible. Since the Ethernet cable has eight conductors (four pairs), not all of them are used. However, it is customary to crimp all of them into the connectors for durability.

Notice that the individual pairs are twisted. This is essential. In conjunction with differential transmission, it eliminates interference, crosstalk, and noise. Where the cables are crimped into the connectors, this twisting is eliminated. Therefore it is desirable to keep these untwisted segments as short as possible.

Rather than making a crossover cable, a stand-alone Ethernet hub can be used, in which case straight-through Ethernet cables connect each device to the hub. If at any time the connection fails, the first thing to check is that the hub is getting AC power, indicated by an LED. The simplest thing to do is to locate the devices close together and use a factory-made Ethernet cross-over patch cord.

National Instruments

National Instruments, founded in 1976 by three computer specialists working in a Texas garage, has become a multinational corporation. After working on a fuel-pump credit card system and various other projects including a waveform generator for the US Navy, the company focused in 1983 on the new Macintosh graphical interface. The end product was LabVIEW, released in 1986.

Laboratory Virtual Instrument Engineering Workbench (LabVIEW) is cross-platform, working in Microsoft Windows and multiple versions of Linux, Unix, and MacOS. LabVIEW is a key component in many types of data acquisition, test auto-mation, instrument control, signal processing and analysis, industrial control, and embedded system design.

The graphical interface has been highly successful. Rather than composing numerous lines of textual coding, the engineer simply uses a computer mouse to draw wires connecting circuit nodes in an intuitive graphical environment. National Instruments offers ample documentation, including tutorials with block diagrams and other graphics that make the processes accessible for users at various levels of expertise.

For oscilloscope connectivity, National Instruments provides this information:

Complete the following steps to connect to and acquire data from a Tektronix TDS3054B oscilloscope:

Download and install the IVI drivers for the scope from the Instrument Driver Network on ni.com. Connect the oscilloscope to your PC using a USB, GPIB, Serial, or Ethernet connection.

Turn on the oscilloscope. In LabVIEW SignalExpress, select File»New Project to create a new project. Press Add Step>Acquire Signals>IVI Scope Acquire to add the IVI Scope Acquire step to the Project View. The Step Setup tab displayscon-figuration options for the step.

On the Configuration page of the Step Setup tab, select Create New from the IVI session name drop-down menu to display the Create New IVI Session dialog box.

In the IVI session name text box, enter TekTDS3054B. This becomes the IVI Logical Name for the session. From the Resource descriptor pull-down menu, select the resource descriptor that specifies the interface and the address of the device. For example, if you are using GPIB to connect to the scope, the resource descriptor might appear as GPIB0::1::INSTR. From the Instrument driver pull-down menu, select tkds30xx, the device driver you downloaded. Click the OK button to close the dialog box and initialize the device. The Vertical and Horizontal sections of the Configuration page update with default settings. From the Channels list, select CH1 and verify that the Enable channel button is set to ON. Update the remaining settings on the Configuration tab to optimize acquisi-tion of your signal. For example, update Range (V) to match the range the oscil-loscope uses for the channel. Click the down arrow on the Run button and select Run Continuously to continuously acquire data from the oscilloscope. Switch to the Data View tab. Drag the tkds30xx(CH1) output of the IVI Scope Acquire step to the Data View tab to view the data you are acquiring.

MATLAB

Mathworks is an American corporation that has released about 100 products including Simulink, Polyspace, Sim Events, and StateFlow. It is best known for MATLAB, a very successful program that like LabVIEW enables users to connect instrumentation such as oscilloscopes, so that waveforms may be remotely analyzed, further processed and saved, as shown in Fig. 8.1.

To this end, MATLAB makes use of numerical computing, involving matrices and algorithms in order to plot data and functions and display them on the computer with highly intuitive user interfaces. MATLAB interacts easily with compatible programming languages including C, C++, Java, Fortran, and Python.

MATLAB has enjoyed outstanding success with engineers, researchers and advanced students since its release in 1984, in part because of its comprehensive documentation. On its website, Mathworks offers an array of tutorials and trainings sessions that are designed to familiarize users with MATLAB's massive functionality.

Onramp is a good introduction. It is a two-hour online course that is designed to familiarize the user with the MATLAB language. Demonstrating the fundamental command structure, the course shows how data are imported, displayed and put to work in a computer that is separate from and augments the oscilloscope or other

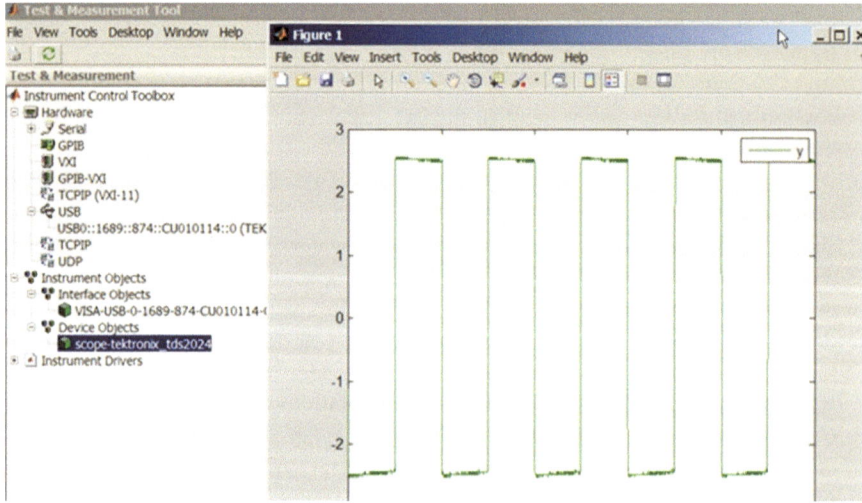

Fig. 8.1 Oscilloscope waveform is shown on a computer via MATLAB Instrument Control Box. (MATLAB)

instruments to which it is connected. As an example, light from a star is accessed and analyzed to determine whether it is moving towards or away from the earth.

MATLAB's web-based and desktop versions are both written in the same language but differ in some respects. Onramp exists in the desktop version. The material in this course is complex, but it is divided into manageable units, beginning with a demonstration of a window in which the user enters commands following MATLAB prompts (>>), then pressing Enter. A star (*) denotes multiplication of scalar numbers.

To enable Computer Vision, MATLAB in conjunction with Simulink and associated algorithms acquires graphics and/or video from imaging hardware. Graphical tools contained in these two separate programs then work together to accomplish the desired visualization and processing. There are also libraries of pre-existing algorithms that permit users to embed designs in hardware that is being developed.

MATLAB and Simulink capabilities for signal processing acquire, measure, and process this data. Users preprocess and filter signals and then analyze them to enable machine learning. The object of the exercise is to detect trends and patterns in connected instrumentation outputs.

Robotics and artificial intelligence are gaining relevance for researchers and engineers, and there is every reason to believe that the trend will continue. System modeling along with new algorithms and autonomous code generation will emerge as MATLAB, Simulink, and similar tools that permit designers to create independent robotic operating systems that will work with a variety of computer operating systems and with an ever-growing variety of sensors and actuators.

LXI

LAN Extension for Instrumentation (LXI), conceived in 2005 by Keysight and VTI Instruments, uses web server technology and Ethernet to enable test and measurement instrumentation applications. Rather than using specific drivers for programmable instruments, users can implement SCPI commands, which make simple bench-top use feasible. There are several advantages in this arrangement. The LAN interface, which can be wireless, works in a peer-to-peer or master-slave configuration. Moreover, triggered events and time-stamped actions are permitted. Multiple test instruments can be included in the network, made possible by decentralized differential signaling.

LeCroy also has a lot of information on this subject. To configure an LXI-compliant LeCroy oscilloscope for LXI connectivity, begin by going to Utilities > Utilities Setup. Then select the Remote tab and press the LXI or LXI11 button.

LXI software is often supplied by the manufacturer, but the quality is sometimes deficient. Since LXI is an open standard, many users choose to create their own software. Linux and MacOS command lines are easy to use. LXI command protocol

*is conventional text compatible with a TCP connection. Here's an example, using
the Telnet command line tool to connect to a Rigol DSA815 oscilloscope:*

```
$ telnet 192.168.1.126 5555
Trying 192.168.1.126...
Connected to 192.168.1.126.
Escape character is '^]'.
*IDN?
  Rigol Technologies,DSA815,DSA8A154402661,00.01.07.00.01
```

```
        Here is a frequency response plot from 100 kHz to
10 MHz using the internal tracking generator:
```

```
# enable tracking generator output
:OUTP 1
# average 1000 measurements
:TRACE:AVER:COUN 1000
# frequency range from 0 to 10 MHz
:SENSe:FREQ:START 10000
:SENSe:FREQ:STOP 10000000
```

Pitfalls

In implementing LXI, there is the potential for encountering problems. Most of
them are easily resolved, although it is possible to have a hardware difficulty that
requires extensive component substitution. That can get expensive, but the good
news is that most problems are either in the software or they are procedural.

One common fault is that on the front panel of an LXI device the LAN status
indicator reads FAULT. Possible causes are a faulty LAN cable or a faulty termination. Ohmmeter readings are not definitive because there may appear to be electrical
continuity while the signal path is problematic. In a long cable run, media can be
pinched due to over-tightened support hardware, causing an impedance mismatch.
The answer is to replace the cable with a known good piece.

Check that the switch or router is ON.

The LAN status will read FAULT if the LXI device is trying to use the wrong IP
address. If the Discovery Tool finds the LXI device but can't access the web page, it
is possible that the IP address does not match the subnet of the computer. The solution is to activate the LCI mechanism (LAN reset). This should align the LXI device
IP address and the computer subnet. If that does not work, an alternate solution may
be to configure the computer to use DHCP or to use a static IP address and subnet
mask matching the subnet of the LXI device.

Ping

Ping is a cross-platform computer utility that permits the user to send a small data packet to a specified domain or Internet Provider address. Resembling a radar transmission, the packet if successful bounces back to the sender's terminal and shows up on the computer screen as a line in the Ping command window along with time of transit.

Ping denotes Packet Internet Groper, and it resembles the Sonar signal that is bounced off submarines and other underwater objects to determine their direction and distance.

Using a computer, you can Ping a digital storage oscilloscope that has an internal server with an IP address. Then, several useful diagnostic tests can be performed. To try out the Ping utility, first ascertain the oscilloscope's IP address, which is determined and stored internally. If this IP address doesn't suit you, you can change it just like a password. To find the IP address on a typical digital storage oscilloscope, press the Utility button. Then press the soft key associated with Utility Page. In the vertical submenu that appears, use Multipurpose Knob a to select I/O. In the new horizontal menu, press the soft key associated with Network Configuration. Then in the vertical submenu that appears, press the soft key associated with Ethernet and LXI. In the window that comes up under LAN Settings, you will find the oscilloscope IP address. It is good to write it down and retain it so that in the future you won't have to interrupt a task to retrieve it.

The Ping operation will not work unless the oscilloscope is connected to your Local Area Network (LAN). To make this connection, plug one end of an Ethernet Cable into the Ethernet port on the rear panel of the oscilloscope and the other end to an Ethernet port that is in your IP cable (or satellite dish) modem. (If there is not a spare port, create one by plugging in an Ethernet hub, switch or router.) You now have a viable Internet connection on your LAN for the oscilloscope and computer. What remains is to configure the computer so that it can Ping the oscilloscope.

To access the Ping utility on a Mac computer (Windows is similar), go to Applications>Network Utility. Then click on Ping. Once you have accessed the terminal window, it is a good idea to bookmark it for future reference. In the terminal, type in the IP address, set the desired number of continuous Pings and click Ping.

The specified number of Pings are performed, the results of each reported in a separate command line in your terminal. If Ping Send To lines read No Route to Host and Host is Down, as shown in Fig. 8.2, then you have to troubleshoot the connection. Most likely the oscilloscope is powered down or the Ethernet cable is not properly terminated at both ends.

If the Ping command lines indicate the round-trip time, as shown in Fig. 8.3, that indicates the operation succeeded. The time in transit of each Ping where you are not going beyond your LAN is a very few microseconds, with a little variation between each event.

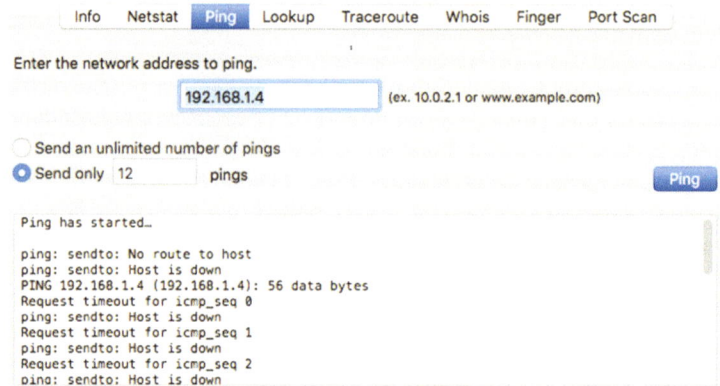

Fig. 8.2 Ping report No Route to Host and Host is Down. (Wikipedia)

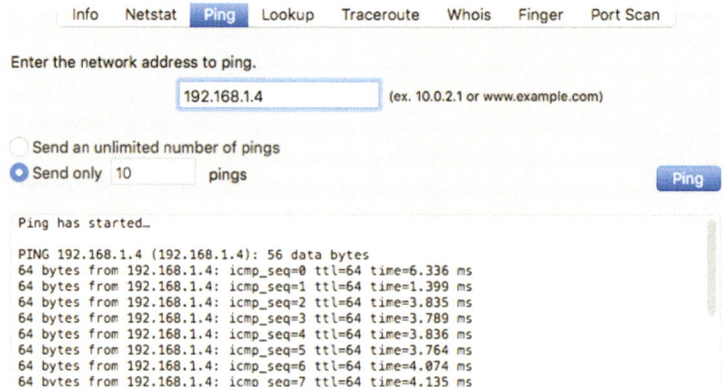

Fig. 8.3 The connection has been successfully Pinged. (Wikipedia)

Ethernet Revisited

Ethernet is the medium of choice in a great many applications, particularly oscillo-scope networking. It is a very good thing to know what is going on behind the scenes, so a closer look is in order. Then we'll discuss Wi-Fi, which is closely related to Ethernet.

Ethernet is relevant to the oscilloscope for two reasons: First, the oscilloscope is the first and obvious choice for troubleshooting existing Ethernet systems and designing new ones. Secondly, the oscilloscope frequently uses Ethernet to connect to a modem, printer, or other oscilloscope.

Ethernet is a frame-based technology. What this means is that the two common Ethernet versions, 10BASE-T and 100BASE-TX propagate data in separate frames, each of which contains seven fields:

The Preamble, which is seven bytes long, consists of alternating ones and zeros to provide synchronization.

The Start-Of-Frame Delimiter consists of one byte with alternating ones and zeros, with two ones at the end.

Each Ethernet node is assigned a unique MAC address, which specifies the *source* of each data packet. These addresses are six bytes in length. They are transmitted in most-significant to least-significant bit order.

Each Ethernet node is assigned a unique MAC address, which specifies the *destination* of each data packet. These addresses are six bytes in length. They are transmitted in most-significant to least-significant bit order.

Length/Type is a two-byte value. If it is less than 1500, it equals the number of data bytes in the data field. If it is greater than 1536, it specifies the protocol that is in the Ethernet payload.

The Data Packet contains 46 to 1500 bytes. If it is less than 46 bytes, it is padded to make it 46 bytes long.

The Frame Check Sequence is a 32-bit cyclic redundancy check, and it determines the validity of the Destination Address, Source Address, Length/Type, and Data fields.

After each frame has been sent, transmitters send at least 12 bytes of idle characters before transmitting the next frame, or they must be idle for that amount of time.

The original 10BASE5 medium was coaxial cable, but now we use Unshielded Twisted Pair (UTP), typically Cat 5e and Cat 6. The latter has a higher twist rate. In conjunction with differential transmission, the twisted conductors inhibit noise, electromagnetic interference, and crosstalk. Point-to-point links are connected by Ethernet repeaters and switches.

Each network node communicates by means of a switch, which forwards successive frames to the destination. Collisions take place only when station and switch talk simultaneously, and data collisions remain localized. The 10BASE-T standard employs full duplex operation. Switch and station can receive and send simultaneously, so contemporary Ethernet is entirely free of collisions. *IEEE 802.3 defines the physical layer and data link layer's Media Access Control (MAC) of wired Ethernet. A number of these standards were promulgated over the years, beginning in 1973 with Experimental Ethernet. The most recent, a work in progress, is 802.3 cm, 400 Gbits/s over multimode fiber. In this time span, IEEE has issued 71 Ethernet standards, covering many Ethernet types and applications. These are denoted in the standards by means of a combination of letters and numbers. Two familiar examples are 10Base-T and 100Base-T. The number (10, 100 or 1000) refers to the propagation speed in megabits per second. The word describes the type of transmission. Base means baseband and broad means broadband.*

The final letter indicates the cable type. T stands for twisted pairs. Added numbers denote the number of twisted pairs. In 100Base-T4, there are four twisted pairs.

If you are using Ethernet cable such as Cat 5e to connect the oscilloscope and computer, it has to be a crossover cable rather than a standard cable. You can purchase a pre-made crossover cable or, if the distance is long, say from one room to another, you can obtain the correct length of cable and put your own male connectors on the two ends, using an inexpensive Ethernet crimping tool and the instructions that come with it in order to create a crossover cable. Both ends should be marked with an X so that in the future, someone does not attempt to use it as a standard Ethernet cable.

If you used a standard Ethernet configuration at each of the two ends, you would be connecting the two transmit pins together and the two receive pins together, which would not work. You need to connect the transmit pins to the receive pins. Standard configurations at both ends are used when you are going from a modem to a computer, or when there is an Ethernet hub at one end. The crossing over is taken care of in the hub or modem. Going from device to device such as in the oscilloscope to computer networking project, we are discussing, or when connecting two computers, the crossover configuration is needed at one end (not both ends, or you are back where you started from). The crossover configuration connects the 1,2 pair on one end to the 3,6 pair on the other end.

Notice that the individual pairs are twisted. This is essential. In conjunction with differential transmission, it eliminates interference, crosstalk, and noise. Where the cables are crimped into the connectors, this twisting is eliminated. Therefore it is desirable to keep these untwisted segments as short as possible.

Rather than making a crossover cable, a stand-alone Ethernet hub can be used, in which case straight-through Ethernet cables connect each device to the hub. If at any time the connection fails, the first thing to check is that the hub is getting AC power, indicated by an LED. The simplest thing to do is to locate the devices close together and use a factory-made Ethernet cross-over patch cord.

Displaying an External Video Feed

As we have seen, the oscilloscope can network to various external instrumentation and display data acquired in that way. A surprising capability of the oscilloscope is its ability to display an actual, moving video signal from outside the instrument, fed into an analog channel through a BNC cable.

To prepare the oscilloscope, begin by pressing the Test button. The horizontal Test menu appears below the display. Press the soft key associated with Application, and in the vertical menu at the left of the display, scroll to Video Picture. This brings up the horizontal Video Picture menu below the display. Use the soft key associated with Display to toggle it on. Notice that the video screen appears at the center of the oscilloscope display. Of course there is still no actual video, but you see the effect of circuit noise, which resembles snow in an analog TV picture when the automatic gain control goes high because the set is not tuned to a transmitted broadcast.

The next menu selection permits the user to toggle between the two TV broadcast standards, NTSC and PAL. NTSC (National Television System Committee) is the analog color system used in North America. It was adopted in 1954 and used continuously until eclipsed by the coming of digital transmission at the start of the twenty-first century. PAL (Phase Alternating Line) is a competing analog TV color encoding system used in countries broadcasting at 625-line/50 field (25 frames) per second.

For our demonstration of video picture display in an oscilloscope, we will use NTSC since that will be the format of our video source. Accordingly, Standard should be toggled to NTSC and left there. Be sure that Channel is selected appropriately.

Since contemporary TV is digital rather than analog, we will have to look off-air for an NTSC signal. An excellent source for an NTSC video signal is an old tape-type VCR. Plug a BNC cable into the selected channel and adapt the other end to pick up the amplified video signal from the VCR. Press Autoset to resize the video window. Contrast and brightness can be adjusted by means of Multipurpose Knobs a and b.

Connecting a Printer and Keyboard

The screen image can be conveyed directly to a printer, which can make one or more hard copies. Either a PictBridge printer or a non-PictBridge printer can be used. To connect a non-PictBridge printer to the oscilloscope, run a USB cable from a USB port on the front or rear panel of the oscilloscope. For a PictBridge printer, use the USB port on the rear panel of the oscilloscope. *In oscilloscopes that do not support PictBridge,* connect a networked printer through the Ethernet port.

To configure the oscilloscope, press the utility button below the display. In the horizontal Utility menu, press the soft key associated with Utility Page and in the vertical menu to the left, use Multipurpose Knob a to scroll to Print Setup.

Press the soft key associated with Select Printer if you are changing from the default printer. Turn Multipurpose Knob a to scroll through the list of available printers. Press the soft key associated with Select Printer to choose the desired printer. To add a non-PictBridge USB printer to the list, plug the printer into a USB host port. The oscilloscope will automatically recognize most printers.

Pressing the soft key associated with Orientation opens the vertical Print Orientation menu so that Portrait or Landscape may be chosen. The Ink Saver mode converts a white trace on a black background to black on white so that large amounts of ink are not expended. PictBridge Printer Settings opens a vertical menu on the right so that paper size, image size, paper type, print date, print name, print quality, and abort print may be chosen.

Many of the oscilloscope functions require letters or numerals to be typed into on-screen fields. For this purpose a very small on-screen keyboard is provided. It is slow and difficult to use. A good alternative is to plug a standard USB Windows keyboard into a USB slot in the oscilloscope. The forward and back arrows position your lettering within the fields.

Electrocardiography

The electrocardiograph is an oscilloscope variation that is built to monitor cardiac activity. A large number of sensors are required for electrocardiograph inputs. This instrument performs electrocardiography (ECG or EKG), a procedure that records electrical (hence physical) activity of the heart by means of external (occasionally internal) sensors affixed to a patient's body. A flat screen (formerly CRT) plots cardiac signals on Y (amplitude)- and X (time)-axes, as shown in Fig. 8.4.

ECG's are usually on wheeled cars so that they can be moved to patient locations as needed. To facilitate monitoring, they are networked to the nurses' station. The connection is usually wireless, and additionally there are loud audible alarms, which sound in the patient's room and in the nurses' station in the event of abnormal cardiac activity.

ECG's display cardiac waveforms, which are composite traces based on electrical signals from the many sensors affixed to the patient. At the electrodes, the voltage levels are very low, in the microvolt range. Each electrode has a single conductor that terminates at the ECG. A pair of electrodes, spaced apart from one another, detect a voltage differential. Due to the complex cardiac waveform, several pairs of electrodes are placed at different locations on the patient's body. Generally, there are twelve electrodes. Ten of them, five pairs, are affixed to the patient's limbs and chest. A highly conductive material that is supplied with the electrode is worked into the skin at each location.

Because of the increased conductivity, ECG's should not be attached to a bench-powered oscilloscope. It is possible a fault current could disrupt the subject's

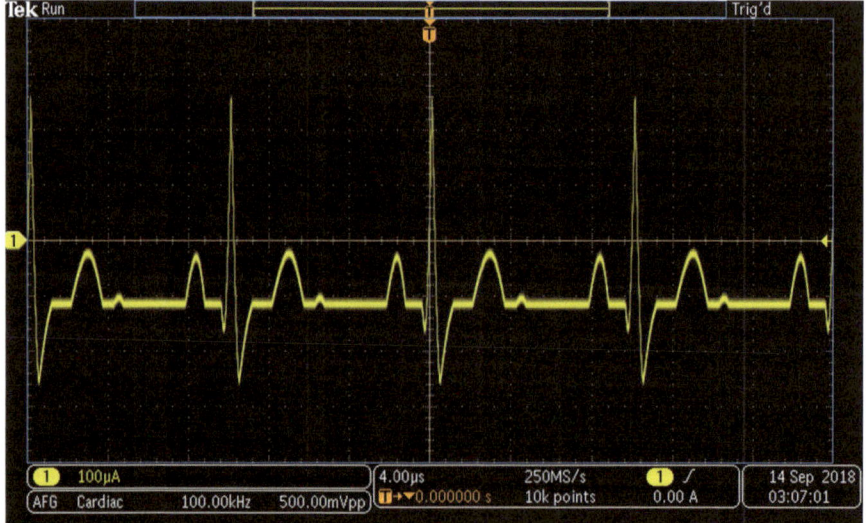

Fig. 8.4 Cardiac waveform. (Author's screenshot)

heartbeat. Professional ECGs are equipped with multiple safeguards that isolate the electrodes from mains power in the event of a malfunction.

Throughout the cardiac monitoring process, the heart's electrical potentials are measured from various angles. They are recorded as 10-second events, and the cardiac waveform is synthesized from this electrical data. The waveform may be found to be normal, or it may display irregularities, which convey information regarding rhythm and overall structure, size and position of heart structure, size and position of heart chambers and heart muscle damage, or conduction system dysfunction. Also, the effectiveness of an implanted pacemaker and administered heart drugs may be ascertained.

Wireless Oscilloscope Technology

There are a variety of wireless oscilloscopes available in the $200 neighborhood. They are mostly single-channel instruments that are built into a probe body with a tip and ground return lead. Wireless oscilloscopes communicate with Apple smartphones and tablets, an Android device or a desktop computer. Essentially, all you need is a wireless equipped phone with a screen and compatible software. We'll take a closer look at these convenient tools, but first here's some background information on wireless technology in general:

In the post Marconi era, early twentieth century, wireless implied radio transmission. The information flew through the air or in a vacuum at virtually the speed of light, unencumbered by cable. Later, the word fell out of favor until it was given new life with the introduction of the TV wireless remote, which uses infrared radiation for line-of-sight control of television programming, volume, brightness, and so on.

Today, in home and office and on the road, wireless technology is widely used as a convenient alternative to Ethernet, Coax, and other cable systems. There are a number of wireless protocols with different characteristics including distance and power limitations, typical applications, and equipment requirements.

Bluetooth is suitable for conveying data over short distances, such as the wireless oscilloscope application. It uses UHF radio waves from 2.400 to 2.485 GHz, and it works for fixed and mobile devices. To market Bluetooth devices, manufacturers must comply with Bluetooth standards and licensing requirements.

Bluetooth is a short-link, frequency-hopping technology. It is a packet-based protocol with master-slave architecture. A single master talks to a maximum of seven slaves, and it supplies the clock. Master and slave, as in many serial protocols, can exchange roles, as needed to initiate two-way communication. Line of sight is not necessary since Bluetooth uses a broadcast communications system. Some Bluetooth applications are:

Communication between a mobile phone and hands-free headset
Control and communication between a mobile phone and car radio. Wireless control
    with iOS and Android phones, tablets, and wireless speakers

Wireless PC networking Wireless communication between PC, mouse, keyboard, and printer

Replacement of RS232 links in GPS receivers, test equipment, bar code scanners, medical equipment, and traffic lights

Wi-Fi is similar but not identical to Bluetooth. It is also closely related to Ethernet. Wi-Fi-enabled devices include desktop and laptop computers, smartphones, tablets, smart TVs, and recently manufactured printers. Wi-Fi communication is LAN-based and permits connection to the Internet.

Wi-Fi uses the 2.4 GHz and 5.8 GHz radio bands, each divided into several channels, which can be time-shared by two or more networks.

Another instance where wireless and oscilloscope technologies intersect is when there is reason to access a signal from high-voltage, high available current electrical equipment that is inside an enclosure that cannot be left open and in which space is limited. The best answer might well be to incorporate the measuring system within the enclosure and make a wireless connection to the outside world, for example, an Ipad.

Of course one would have to get the data out through the heavy grounded steel enclosure. An RS-232 connector on the front panel would suffice. (RS-232 is still often used in industrial machinery.)

Returning to the wireless oscilloscope probe, with its 200-foot range, Aeroscope employs Bluetooth in its 4.1 model, as shown in Fig. 8.5. For displaying waveforms, it uses an Android tablet or Ipad.

Networking oscilloscopes with each other and with local and remote computers. This can be accomplished, the easiest being by loading a waveform with or without settings into a USB device and moving it to a nearby computer. Additionally, oscilloscope and PC or Mac can be connected over the Internet using proprietary software, so that the oscilloscope can be remotely viewed and controlled. Methods are shown.

Fig. 8.5 Aeroscope
wireless oscilloscope probe

The analog bandwidth is 100 MHz with a sample rate of 500 Ms/s.

The input range is +40 volts to −40 volts with a DC accuracy of ± 3 percent. The sample memory depth is 10 K. Input impedances are 10 Megohm and 12 pF. Resolution is 20 mV/division to 10 V/division. From Digikey, the price is $199.

Another well-designed and highly regarded wireless oscilloscope the IkaScope, shown in Fig. 8.6, has substantial specifications.

The 30 MHz bandwidth is more than adequate for the type of measurements contemplated as is the 2000 MS/s sampling rate. For a small hand-held instrument, the input range, 80 V_{pp} is outstanding. The user interface is shown in Fig. 8.7. With AC/DC coupling, 4000 pts memory and 200 FPS maximum refresh rate, the 359 Euro price including taxes and with free world-wide shipping is not excessive.

Fig. 8.6 IkaScope wireless oscilloscope

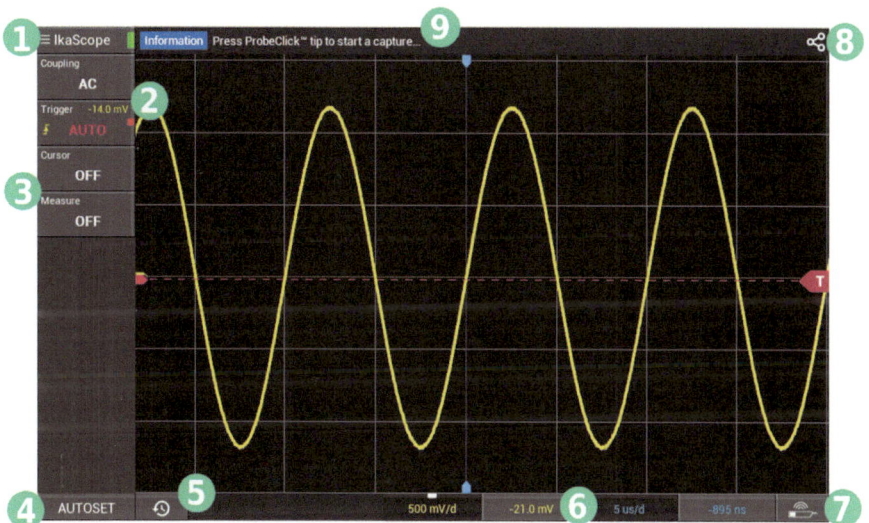

Fig. 8.7 IkaScope User Interface displayed on an oscilloscope

The internal battery is recharged through a micro-USB charging cable. Even while charging, the instrument is isolated from ground potential, ensuring safe measurements on grounded circuits. IkaScope displays in desktop computers with Windows, Mac or Linux operating systems, as well as in tablets and smartphones.

Chapter 9
PC-Based Oscilloscopes

Abstract The PC-based oscilloscope, also known as USB-based oscilloscope, consists of a compact module that is connected to a user-supplied laptop or desk computer, which, with manufacturer-supplied software, provides processing and display. The module is far less expensive than a stand-alone bench-type oscilloscope with equivalent features and specifications. Major oscilloscope manufacturers, arbitrary function generators, and types of waveforms are also discussed.

Hardware configurations vary, but what PC-based oscilloscopes have in common is that they consist of a separate module that is connected to a user-supplied computer for the purpose of displaying, measuring, and analyzing waveforms. These waveforms are derived from electrical signals applied to inputs in the module by means of probes or BNC cables. The sole connection between the module and the computer is a cable, preferably USB, which carries data and control signals in both directions and also supplies electrical power to the module.

These instruments are more properly called USB-based oscilloscopes to emphasize that many of them are now truly cross-platform rather than exclusively Windows compatible and also to suggest fast, reliable USB cabling as the preferred medium.

A large number of manufacturers offer these modules, and the quality varies from minimal hobby-grade boxes to superb instruments with features and specifications approaching those of the most advanced bench-type oscilloscopes. What is typically included with the USB-based oscilloscope is:

- A compact, environmentally sealed module in a rugged enclosure with no external knobs or buttons – just perhaps an LED power light and the number of ports corresponding to the number of channels, usually one or four. The module contains one or more printed circuit boards, consisting of an interface that provides electrical isolation and automatic gain control with the required number of ADC's along with requisite buffer memory and perhaps a portion of the digital signal-processing capability.
- Probes, one for each channel, BNC cables and the USB cable.
- Software, provided by the manufacturer, which can be a USB flash drive, CD, or download available from the manufacturer's website.

© Springer Nature Switzerland AG 2020   217
D. Herres, *Oscilloscopes: A Manual for Students, Engineers, and Scientists*,
https://doi.org/10.1007/978-3-030-53885-9_9

- Also, on the website, downloadable owner's manual and access to data sheets, specifications, tech help, users' forums, videos, white papers, and the like.

Naturally there are some advantages and disadvantages inherent in the USB-based instrument compared to the conventional bench-type oscilloscope. Because the user supplies a pre-existing computer, the USB-based oscilloscope is far less expensive than a bench-type oscilloscope of comparable specifications. Given the fact that almost everyone on the planet owns a computer, its cost is usually not a factor. If you want a dedicated laptop to go with the USB-based oscilloscope, an older, secondhand instrument will work.

In addition to providing the computing power to go with the module, the computer has a large, clear display. Once the acquired waveform is in the computer, it can be copied to available applications such as Microsoft Word or Excel. Custom programs are an option, and the computer cab be networked beyond the LAN. Also, the computer's hard drive has enormous storage capacity. With a laptop, remote and difficult field locations are feasible.

These advantages should be weighed against some palpable drawbacks in the USB-based implementation. Weak-signal resolution depends upon robust shielding to mitigate power supply and electromagnetic noise emanating from computer circuitry. When the USB-based module is streaming to the computer, data transfer ratios can be lower than suggested in the instrument's specifications. This is highly dependent on the cabling method. Use USB cable, as short as possible. The USB-based software is supplied along with the instrument. The user should ascertain in advance whether it is compatible with the computer's operating system.

Outstanding products among USB-based oscilloscopes are Pico Technology's PicoScope 2000, 3000, 4000, 5000, 6000, and 9000 Series instruments, the prices ranging from $115 to $26,000 depending upon bandwidth, number of channels, and other specifications. Here's a rundown:

The PicoScope ultracompact 2000 Series has two or four channels (+ 16 digital with MSO). Bandwidth is 10–100 MHz. Maximum sampling is 16 GS/s. Memory is 8 kS to 128 MS. All models have built-in function generators and AWG's.

The PicoScope 3000 Series has two or four channels (+16 digital WITH MSO). Bandwidth is 50–200 MHz. Maximum sampling is 1 GS/s. Memory is 64 MS to 512 MS. All models have built-in function generators and AWGs.

The PicoScope 4000 Series has two, four, or eight channels. Bandwidth I 5–20 MHz. Maximum sampling is 80 MS/s. Memory is 256 MS. Some models include AWGs, and some models have differential inputs and IEPE.

The PicoScope 5000 Series has two or four channels (+16 digital with MSO). Maximum sampling is 1 GHz/s. Memory is 128 MS to 512 MS.

The PicoScope 6000 Series has four channels. Bandwidth is 250 MHz to 1 GHZ. Maximum sampling is 5 GS/s. Memory is 256 MS to 2 GS.

The PicoScope 9000 Series has two or four channels. Bandwidth is 15–25 GHz. Maximum sampling is 1 MS/s. Memory is 32 kS. Included are 9.5 GHz optical inputs, clock recovery, and differential TDR/TDT.

Earlier we described how to convert a CRT television into an oscilloscope by feeding the signal into the vertical deflection coils. The resulting instrument, if you want to call it that, is very limited in terms of resolution, bandwidth, and features. In fact, the project, while interesting, instructive and inexpensive, is not at all comparable to even the lowest-end manufactured oscilloscope.

Another project in the same category involves building a PC-based oscilloscope using the audio input of a computer. This is generally known as a sound card oscilloscope. Detailed plans are available on the Internet.

The project consists of feeding the signal into the audio input of a PC, where it is processed by the sound card so that it can be displayed. For this to happen, enabling software must be installed in the computer. This software, some of it free of charge, can be downloaded from the Internet.

There are two problems. First, the sound card is capable of displaying frequencies that correspond roughly to the human hearing range, 20 Hz to 15 kHz. Any serious oscilloscope goes way beyond that. And secondly, the computer sound system will tolerate only a very low voltage level. To deal with this problem, the input has to pass through a resistor-diode network, which attenuates the signal to a safe level. Even so, over 12 volts should not be applied to the probe. So don't expect much from this project, but the cost is next to nothing so it might be an instructive diversion if you are so inclined.

Besides TV and sound card oscilloscopes, the Internet has plans for building Arduino, smartphone, and Raspberry Pi-based oscilloscopes.

Major Oscilloscope Manufacturers

The big three in the USA are Tektronix, Teledyne-LeCroy, and Keysight (formerly Agilent and before that Hewlett-Packard). All three make a wide range of instruments having various numbers of channels, different bandwidths, sampling rates, and other specifications. And, of course, the prices vary by orders of magnitude.

The number of channels is one of the determining factors. For each channel, there is a separate probe and set of front-end components, controls, and signal paths all the way to the display.

Bandwidth is a very significant factor in pricing, because at higher frequencies capacitive reactance diminishes and inductive reactance increases. Both of these high-frequency phenomena strongly attenuate the signal because capacitive reactance occurs in parallel and series reactance occurs in series with respect to the signal path. These effects are seen in probes, cabling, internal wiring, and circuit-board traces and connectors and components, notably within solid-state devices. The ADC's are especially price-sensitive with regard to bandwidth.

The big three manufacturers' oscilloscopes have similar interfaces with mostly equivalent controls. The situation is like a Windows user learning to work a Mac computer.

Tektronix makes a superb line of oscilloscopes. The high-quality, but no-nonsense interface is exceptionally user-friendly, and the online and print documentation is clear, concise, and comprehensive. Very advanced probes can be automatically and precisely compensated merely by pressing a soft key. A complete line of well-built accessories contributes to the Tektronix experience.

The Teledyne-LeCroy oscilloscope, with its large, clear display, is a thing of beauty. For users with plenty of bench space, LeCroy oscilloscopes have much to recommend them. The LabMaster offers up to 100GHz, 80 channels, 240 GS/s, and 1536 Mpts/ch. The modular design allows for convenient upgrades.

Keysight continues the high-quality Agilent tradition, with large touchscreen displays, high specifications, and excellent documentation. These oscilloscopes are known for precise signal integrity and outstanding display clarity. Keysight does exceptional research and development and is well-known for its robust product innovation.

These are the big three, manufacturers of superb instruments that are rugged, reliable, and beautiful. Beyond these three, there are several high-quality oscilloscope manufacturers whose products are found in the world's finest laboratories and foremost learning centers.

Rohde & Schwarz makes excellent bench-type oscilloscopes with exceptionally clean signals, low noise floors, and large, clear displays. Bandwidths range from 50 MHz to 6 GHz. Among oscilloscope manufacturers, the prices are modest, including a superb handheld instrument that starts at just over $3,000.

Yokogawa is a major international corporation that manufactures all kinds of specialized industrial test and measurement equipment that is used worldwide. They introduced their first oscilloscope in 1988, and they have been refining the product line since then. The website is a wealth of information, with a huge amount of application notes, manuals, product-specific resources, training modules, and videos. You can access this valuable material at tmi.yokogawa.com.

Rigol manufactures digital oscilloscopes, waveform generators, multimeters, data acquisition instruments, DC power supplies and loads, spectrum analyzers, and RF signal generators. The oscilloscopes are highly competent, well-made instruments. What is interesting to some users is that Rigol offers some models with reduced specifications for remarkably low prices. For example, the DS2102E oscilloscope has two channels, a 100 MHz bandwidth, 50,000 wfms/s capture rate, and 28 Mpt maximum memory depth. The selling price is only $647 USD. Despite reduced specifications, the oscilloscope is well-made with an advanced user interface, and looks like it would last a lifetime. Some users would question the channel count and bandwidth, but for the price you can afford to ask when the last time you used four channels simultaneously or needed to display a signal faster than 100 MHz.

The Rigol is a good oscilloscope, light weight but durable, and it offers a full range of features such as Math, FFT, and the like. And there are some extras. When you push the Help button, on-screen text displays what is relevant to whatever function is currently active. There are some eccentricities – to save a current waveform to a flash drive in the USB slot, you have to press the Print button.

BK Precision, like Rigol, offers some oscilloscope models with lower specifications and prices, while maintaining a high-quality product. The handheld models are durable, digital storage oscilloscopes with attractive interfaces and a full range of features.

Model 2511 is 60 MHz bandwidth, 1 GS/s, with non-isolated inputs. The price is $945.

Model 2512 is 100 MHz bandwidth, 1 GS/s, with non-isolated inputs. The price is $1,145.

Model 2515 is 60 MHz bandwidth, 1 GS/s, with isolated inputs. The price is $2,185.

Model 2516 is 100 MHz bandwidth, 1 GS/s, with isolated inputs. The price is $2,445.

A major plus in the BK Precision instruments is the lengthy owner's manual. It covers product features in depth, and there is copious generic information relevant to all digital storage oscilloscopes. The website also contains lots of information, including a link to BK Precision's YouTube channel.

Types of Oscilloscopes

Analog as an oscilloscope type refers not to the type of signal that can be processed, but to the basic architecture of the instrument and to the way signals are conveyed to the screen. A digital oscilloscope has analog channels, each with an input port. After preprocessing, these signals are fed to the ADC's, where they are sampled and digitized before going on to the memory, processor, and display.

In an analog oscilloscope with CRT, the signal is fed to the vertical deflection plates, and a time-based waveform is fed to the horizontal deflection plates, so that a moving beam traces the waveform on the screen, which is at the relatively large visual output end of a CRT. The analog oscilloscope is considered obsolete and is no longer manufactured.

The digital storage oscilloscope (DSO), which is not encumbered by the heavy and bulky CRT tube, is lighter and there is far less heat to be dissipated, making for an easily portable instrument even in the bench-type models. There are many other advantages in terms of signal measurement, processing, and storage. Also, the flat screen is a big plus. The DSO is divided into several subtypes:

The digital phosphor oscilloscope is analogous to the old analog phosphor oscilloscope. In that instrument, a chemical coating was added to the display so that the trace would persist rather than appear for only an instant. In the digital phosphor oscilloscope, there is no such chemical, but a similar effect is achieved electronically, due to the overall architecture of the processing section.

In the conventional digital storage oscilloscope, a time interval occurs between the end of each scan and the point at which the trigger initiates a new scan. This delay is responsible for the fact that signal activity does not immediately appear in the display. Serial processing is the limiting factor.

The digital phosphor oscilloscope remedies this problem by employing parallel processing in a separate processor. Waveforms are captured and stored without regard to display speed. As in the analog phosphor oscilloscope, the trace as displayed becomes more intense every time the waveform occurs. Besides increased intensity, a greater number of very short-term events are retained in the display, including transients that would otherwise not be seen. Also, unlike a conventional oscilloscope, the digital phosphor oscilloscope provides a real-time intensity Z-axis.

The digital phosphor oscilloscope is not for the faint-hearted. Tektronix offers four models in its 7000 series:

The DPO7054C has a 500 MHz analog bandwidth rating. The sample rate is 5 GS/s – 20 GS/s. The record length is 25 M points to 250 M points. The list price is $19,200.

The DPO7104C has a 1 GHz analog bandwidth rating. The sample rate is 5 GS/s – 20 GS/s. The record length is 25 M points to 250 M points. The list price is $24,000.

The DPO7254C has a 2.5 GHz analog bandwidth rating. The sample rate is 10 GS/s – 40 GS/s. The record length is 25 M points to 500 M points. The list price is $32,300.

The DPO7354C has a 3.5 GHz analog bandwidth rating. The sample rate is 10 GS/s – 40 GS/s. The record length is 25 M points to 500 M points. The list price is $41,300.

All models have four analog channels.

The digital sampling oscilloscope is capable of displaying very high-frequency signals, well beyond the nominal oscilloscope specification. It works on repetitive signals by collecting samples from successive waveforms and combining them to create a displayed image. Accordingly, it is possible to compile a waveform of 50 GHz and higher signals. The downside of this implementation and reason that it is not used except where needed for very high-frequency signals is that the dynamic range is greatly reduced. In fact, the digital sampling oscilloscope cannot be used for signals in excess of 3 Volts peak-to-peak, and it is not feasible to use diode protection because this would affect the frequency response.

The mixed-domain oscilloscope (MDO) is normally used as a conventional oscilloscope. The analog channels permit signals to be displayed, measured, analyzed, and stored in the time domain. Also, there may be MSO-capable digital inputs. When placed in the Math mode and the soft key corresponding to FFT is pressed, the MDO oscilloscope displays both the time-domain signal that is applied to an active channel input, and simultaneously, in split-screen format, its frequency-domain version.

In contrast, the mixed-*signal* oscilloscope (MSO) is capable of displaying in time domain one or more analog signals and also one or more digital signals simultaneously. Suppose you have, in design or prototype stage, a new model whose operation is to be evaluated under varying loads and conditions or, as another example, an existing device whose digital output has become intermittently buggy.

One way to proceed would be to connect the output of the DC power supply to an analog channel and the digital outputs to multiple logic channels. Then, in MSO mode, observe the two displays as shown in the split-screen format and ascertain whether the digital output is exhibiting problems that correlate to power-supply fluctuations. (You could apply mild heat or vibration to power-supply components while watching the digital output.) This diagnostic procedure could be the basis for a simple parts substitution that would restore normal operation.

Handheld Oscilloscope

Most oscilloscope manufacturers offer a range of handheld, battery-powered models. Because of size and weight constraints, they tend to be modest instruments compared to bench-type instruments, but many of them have impressive arrays of features and reasonable bandwidths for most applications. Most handheld oscilloscopes of necessity have displays that are less impressive than bench-type oscilloscopes, but you quickly learn to live with that in the context of convenience and ease of handling.

Besides lower cost and portability, most handheld oscilloscopes have analog channel inputs that are isolated/insulated from ground and from one another (check user's manual). This is a great advantage because it means that they are not subject to the great hazard involved in the use of an AC-powered, bench-type oscilloscope. As emphasized in the Introduction and throughout this book, the ground reference lead or equivalent conductor in a cable connection must never be connected in the equipment being tested to any wire, terminal, or conductor that is referenced to but floats above ground potential. To make such a connection will cause a sudden high-level fault current to flow through parts of the equipment being tested, the oscilloscope and grounding conductor of the electrical supply, back to the electrical service and system ground.

The handheld, battery powered oscilloscope bypasses this hazard because it is not connected to the facility electrical supply and grounding system. The equipment under study can be safely probed without regard to polarity or grounding.

That being said, it is absolutely essential to observe voltage limits and Cat ratings, which are usually printed on the oscilloscope enclosure, noted in the specifications and explained in detail in the user's manual. All electricians, electronics technicians, and electrical and electronics engineers should have a thorough understanding of Cat ratings. They are explained in depth on the Internet. Some user's manuals, especially those that go with multimeters, have good explanations.

Another advantage of the handheld oscilloscope is its light weight and ruggedized case. If dropped from a height of five feet onto a concrete floor, it should not be damaged unless due to an irregular surface something cracks the recessed screen. The handheld instrument is relatively impervious to small amounts of moisture encountered outdoors or dust in a factory setting.

Fluke ScopeMeter

Fluke has manufactured high-quality electrical test equipment for nearly 70 years. A leading product today is the superb Fluke ScopeMeter. A little larger than most handheld oscilloscopes, it combines the functions of multimeter and oscilloscope in a single instrument with ruggedized case, rechargeable battery, and advanced oscilloscope capabilities such as Math and FFT. The top of the sine ScopeMeter 190-504/s sells for $6000 and up.

The oscilloscope display and multimeter readout are remarkably clear and controls and documentation are absolutely user-friendly. There are four electrically isolated inputs, Cat III 1000 Volt/Cat IV 600 Volt rated. Buyers can choose from 60 MHz, 100 MHz, 200 MHz, and 500 MHz bandwidth models. Depending on model and channels used, the sampling rate is up to 5 GS/s with up to 200 ps resolution. The memory is 10,000 samples per channel waveform capture, permitting the user to zoom in on details. Additionally, there is a 999 count digital multimeter.

The ScopeMeter incorporates a feature known as Trendplot, which is valuable in finding intermittent faults. It creates a paperless, electronic record of voltages, amps, temperature, frequency, and phase for all four inputs for up to 22 days with time and date stamp.

Another function, ScopeRecord, stores in memory up to 30,000 data points per channel to capture intermittents and fast glitches. These high-resolution waveforms can be recorded for up to 48 hours and stored in memory. With stop-on-trigger mode, the ScopeMeter 190-504/S automatically recognizes a power failure and stores the waveform data preceding it. Using Waveform Zoom the user can look at individual power cycles.

Sources for Signals

Often you need to feed a known signal into an oscilloscope to check out the settings and see how it looks in the display. Or, you may want to inject a known signal at various locations in a piece of electronic equipment, transmission line, or network to check continuity or to see how the signal is modified in its downstream journey. This is the method TV technicians employed to troubleshoot domestic equipment in the days when there was a shop on every street. These same procedures can be generalized today to fix highly specialized industrial equipment such as the user interface for a variable speed drive or a radar installation.

There are many sources for signals, and it is instructive to make use of them and to see which ones are best in specific applications. You can even make your own waveform, using the arbitrary function generator, as we shall see.

The quick and convenient signal source for an oscilloscope is the internal arbitrary function generator. This is a step above the simple waveform generator because it permits the user to create an endless variety of signals for specific applications and for purposes of comparison.

Beginning with the most fundamental, an internal arbitrary function generator typically contains a library of a dozen basic, often-used waveforms: sine, square, pulse, ramp, DC, noise, $sin(x)/x$, Gaussian, Lorentz, exponential rise, exponential decay, Haversine, and cardiac, as shown in Figs. 9.1, 9.1, 9.2, 9.3, 9.4, 9.5, 9.6, 9.7, 9.8, 9.9, 9.10, 9.11, 9.12, and 9.13.

These waveforms in the AFG can be modified. Press AFG to bring up the horizontal AFG menu bar. Then press the soft key associated with Waveform Settings. The vertical Waveform Settings menu appears to the right of the display. Multipurpose Knob a or the keyboard sets frequency. For the sine wave the maximum available frequency is 50.000 MHz. (Square wave is less, 25.000 MHz, due to high-frequency harmonics needed to comprise the square wave.) Notice that at the highest frequency, vertical rise times, horizontal high and low voltage levels and square corners are not realized.

As you adjust frequency, period automatically conforms. Or, you can use Multipurpose Knob b or the keypad to adjust period, and frequency will fall into line. Similarly, pressing the appropriate soft keys, amplitude and offset can be adjusted.

Pressing the soft key at the right, output settings can be changed. For Load Impedance, High Z is the default and should be used most of the time. Fifty-ohm load impedance is used occasionally when necessary to match impedance, but care must be used because it is possible to overload the AFG or distort the measurement.

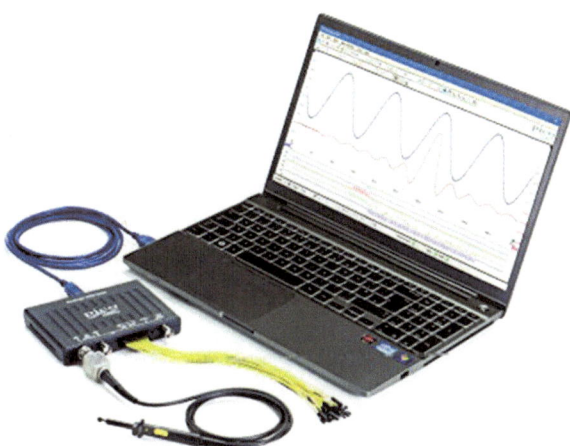

Fig. 9.1 The sine wave is a very basic waveform found throughout nature and in human-made machinery and electronic equipment. It occurs in mathematics, physics, engineering, signal processing, and elsewhere. Plotted o a graph where the vertical Y-axis represents volts and the horizontal X-axis represents time as in the oscilloscope's time-domain mode, the sine wave appears as a smooth periodic curve. All the power is at a single frequency, and it is free of harmonics. The curve has a pleasing appearance because of its symmetry, but the sound can be unpleasant because at moderate volume it tends to overload the auditory receptors at a single frequency. (Author's screenshot)

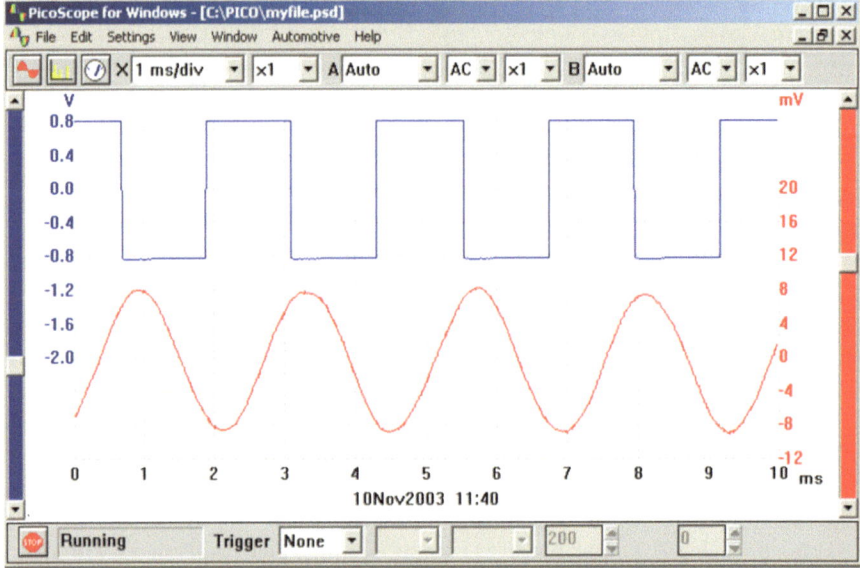

Fig. 9.2 The square wave, a special case of the pulse wave, but with a 50 percent duty cycle, has an amplitude that alternates at a fixed frequency between high and low levels. The square wave has a high harmonic content due to the very fast rise and fall times, which are high-frequency components. For this reason, an AFG is capable of generating a complete square wave only up to a lower maximum frequency compared to a sine wave. The output of a clock circuit is a square wave. (Author's screenshot)

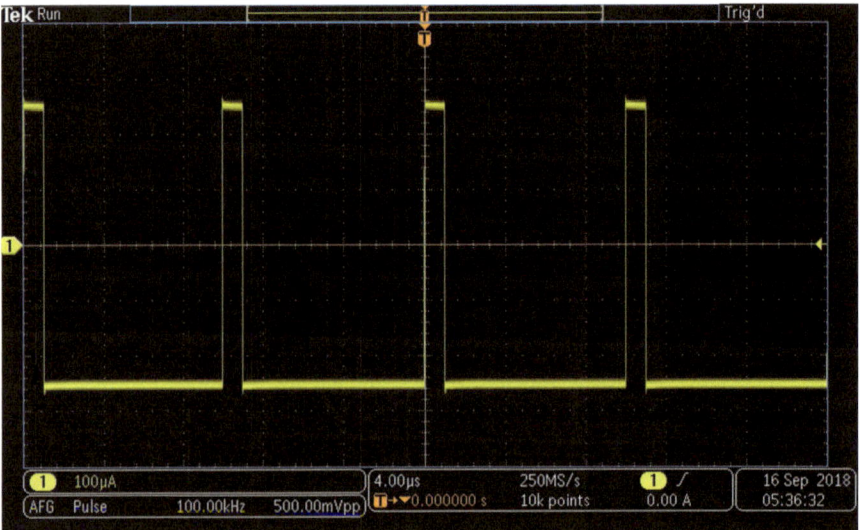

Fig. 9.3 Pulse resembles a square wave, but the duty cycle is not restricted to 50 percent. Pulse-width modulation is one of the methods employed by variable frequency drives to control the speed of AC induction motors. (Author's screenshot)

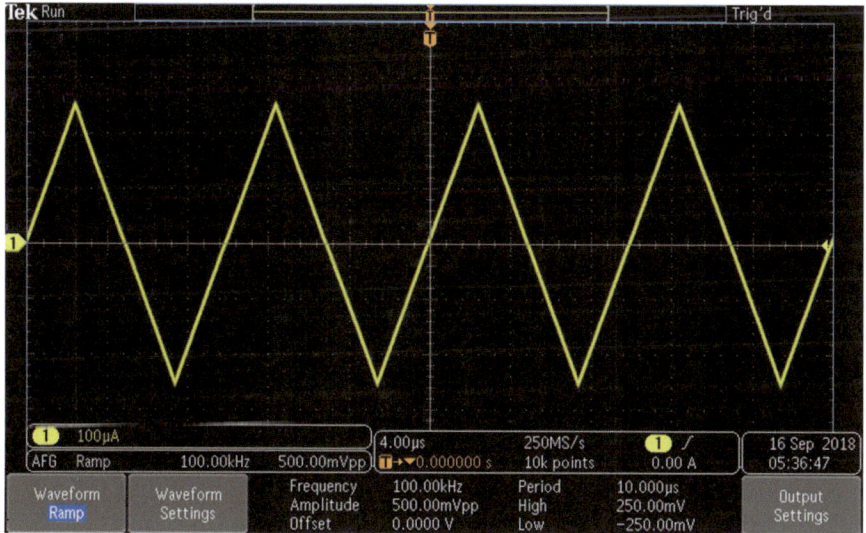

Fig. 9.4 Ramp wave, of which the saw tooth wave is a variation, was used in CRT scanning in analog oscilloscopes and TVs. (Author's screenshot)

Fig. 9.5 DC appears in the oscilloscope display as a straight horizontal line, above or below the X-Axis. AC coupling in an oscilloscope suppresses the relatively high DC component so that ripple in the output of a DC power supply can be displayed and measured. (Author's screenshot)

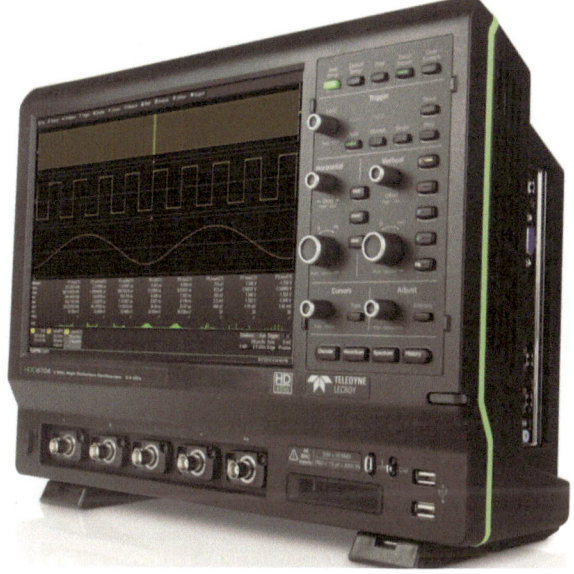

Add Noise is very useful as described earlier. Amounts vary from 0 to 100 percent.

And so we see that in addition to the 13 available waveforms, parameters may be adjusted to create variations.

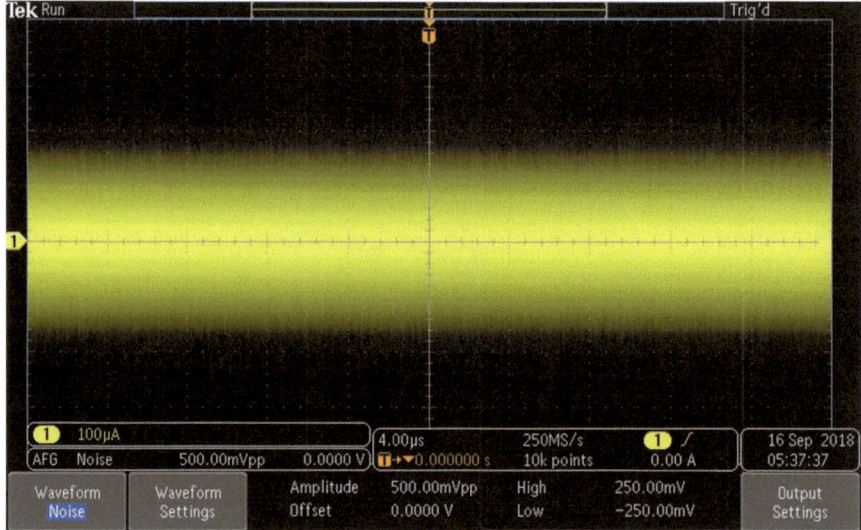

Fig. 9.6 Noise may have specific amplitude but no specific frequency. Varying percentages of noise can be added to other signals in order to evaluate mitigation techniques such as bandwidth limiting and waveform averaging. (Author's screenshot)

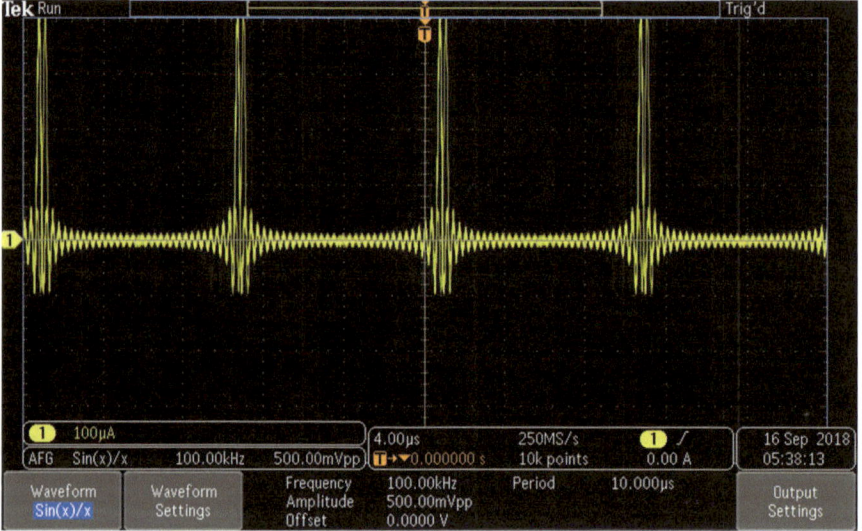

Fig. 9.7 The sin(x)/x provides a ringing pulse useful for testing circuits with inductance. A sharp single pulse will have ringing with an inductive circuit where the higher frequency components are cutoff. This function simulates that behavior. (Author's screenshot)

The AFG also permits the user to create numerous original waveforms. To see how this works, press Default Setup and restore the AFG. Then press the soft key associated with Waveform and use Multipurpose Knob a to scroll to Arbitrary.

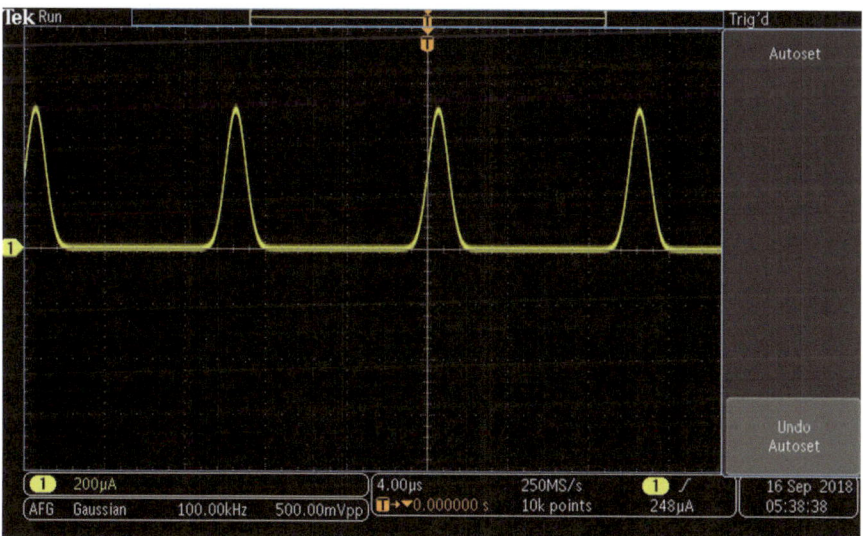

Fig. 9.8 Gaussian (normal) distribution, sometimes known as a bell curve, is a graph showing the number of occurrences of random events under certain frequently encountered circumstances. Physical quantities that are the sum of many independent processes, such as measurement errors, often conform to this curve. (Author's screenshot)

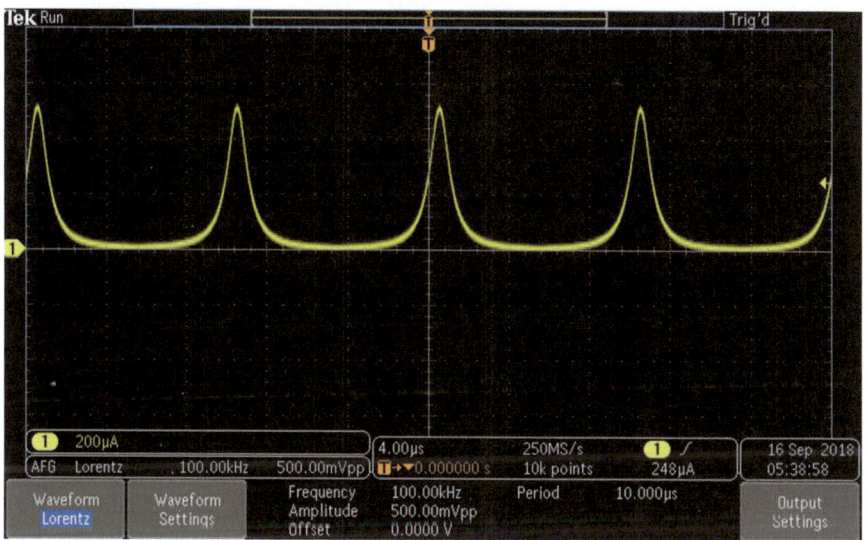

Fig. 9.9 Lorentz curve is similar to the Gaussian curve, but it is more sharply biased. It is often used to represent income distribution. (Author's screenshot)

The AFG can generate up to 131,072 points in an arbitrary waveform. Unique waveforms can be created from any of the four ARB memories, the analog channels, the reference waveforms, the Math waveforms or the 16 digital channel waveforms.

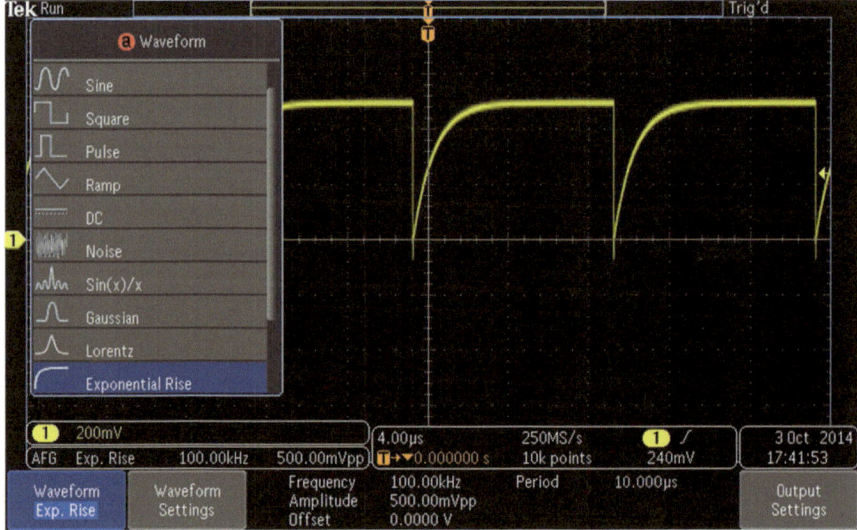

Fig. 9.10 Exponential rise is similar to a saw tooth wave, but the rise is exponential rather than linear. (Author's screenshot)

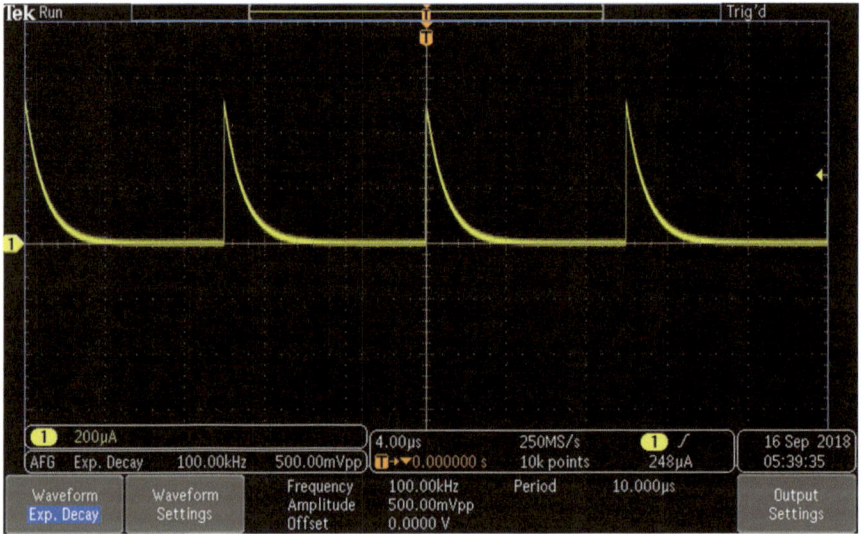

Fig. 9.11 Exponential decay is similar to a saw tooth wave, but the decay is exponential rather than linear. (Author's screenshot)

You can also use a .CSV spreadsheet file stored externally or a predefined template. You can modify the arbitrary waveform by means of an on-screen editor and then replicate it out of the generator. For more complex projects, use Tektronix

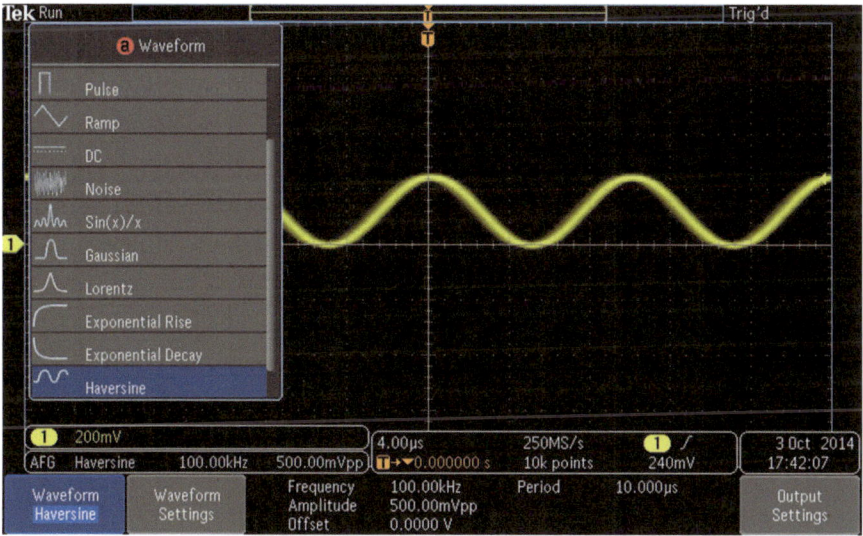

Fig. 9.12 Haversine waveform is sinusoidal but, unlike the sine, it rides above the *X*-Axis, the voltage never going into negative territory. (Author's screenshot)

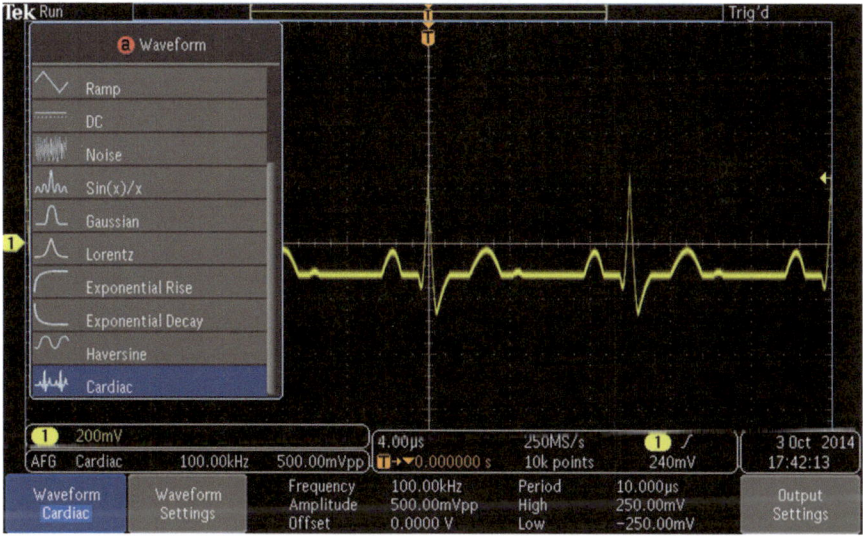

Fig. 9.13 Cardiac waveform is a graph of electrical activity of the heart. (Author's screenshot)

ArbExpress PC-based waveform creation and editing software, available free of charge at www.tektronix.com/software.

To proceed, press the soft key associated with Waveform Edit. The menu that appears allows the user to manipulate existing waveform points, add and delete points and edit the voltage level.

You can use the Edit Waveform menu to create new arbitrary waveforms in the instrument by loading them from files or live channels. Use the vertical Edit Existing menu to change, add or delete points in a current waveform.

When you enable the internal editor, the screen splits into a smaller top window and a larger bottom window. The smaller top part of the screen is an overview, showing the entire waveform memory. A box contains the zoomed-in portion of the waveform. The larger, lower part is the zoomed-in representation of the overview. The lower part shows up to 500 points of the record.

Turn Multipurpose Knob a to select a point to edit. Turn Multipurpose Knob b to set the voltage level of that point. Use the side menu items to add or remove points from the waveform. Press Create New from the lower menu to make a new arbitrary waveform. In the side menu that now appears, turn Multipurpose Knob a or use the keypad to define the number of points in the waveform. You can have up to 131,072 points. Turn Multipurpose Knob b to select the basic function. It can be square, sine, ramp, pulse or noise. Press OK Create to make the new waveform.

Press Load Waveform. Turn Multipurpose Knob a to choose which of the waveforms to display. Then, press OK Load. Connect a BNC cable from Aux Out on the back panel to Channel One input on the front panel. The arbitrary waveform appears in the display. In the lower menu, press Save Waveform. Turn Multipurpose Knob a to select the location where the waveform is to be stored. Press OK Save. The waveform can be named by pressing Edit Labels and using Multipurpose Knob a and the lower menu to assign a name.

The autonomous, bench-type waveform generator is an alternative. It takes up separate bench space, it is an added expense and requires a little extra time to warm up and connect to the device under investigation and to the oscilloscope, but there are numerous capabilities and features that won't be found elsewhere. While the controls and operating procedures are simple and intuitive, a typical user's manual is lengthy only because of the many features that have to be covered.

A stand-alone waveform generator may have as many as four outputs as opposed to the single output of an oscilloscope-integrated AFG. The autonomous instrument, moreover, can function as a logic source, either a pulse generator that synthesizes a stream of square waves or pulses of varying duty cycle or the more complex pattern generator, which provides eight, sixteen, or more synchronized digital pulse streams to be applied to a computer bus or communications network.

The stand-alone waveform generator, moreover, is capable of applying various types of modulation to standard signals, or applying impairments, not just noise as in a typical integrated instrument. Sine and other waveforms can be frequency-swept.

Other useful signals can be extracted from demo boards, supplied by test equipment manufacturers to demonstrate the capabilities of individual or groups of models. These demo boards are often USB-powered and contain a separate terminal for each signal so that hook tip probes can be connected. Sometimes, when boards require high current, Y-type dual USB cables are included to prevent overloading of a single USB port in the host machine.

Demo boards from multiple manufacturers can be viewed in any standard oscilloscope provided voltage limitations at the signal outputs and current limitations at the power inputs are observed. A Rigol demo board, for example, has a very interesting swept-frequency sine wave that doesn't seem to be available elsewhere.

Printed circuit boards can be probed for interesting signals and it is an instructive exercise to figure out how to power and probe them, especially in conjunction with schematics.

To look at the human voice as displayed in the time domain, the appropriate transducer is of course the microphone. This device can be obtained from any old telephone. You have to be aware that there are two types of microphones and to make them work they have to be configured differently. The electret microphone, capacitor-based and hence also known as a condenser microphone, converts the mechanical energy of changes in air pressure to audio-frequency electrical current. All you have to do is probe the two terminals and speak into the microphone to see your voice displayed as a non-periodic trace in the oscilloscope. You will probably have to default the instrument and press Autoset.

The other type of microphone requires a separate external DC power source. An example is the resistive carbon microphone. It consists of granulated carbon behind a thin diaphragm that vibrates in response to acoustic energy in the form of changes in air pressure. Two leads are connected to opposite sides of the carbon and a low voltage DC power source is placed in series. Polarity does not matter. As the carbon compresses and decompresses, its resistance changes, resulting in voltage fluctuations at the oscilloscope's analog input. Here again, press Autoset.

Another source for electrical signals that can be viewed in an oscilloscope is the Arduino Uno development board. This board was first introduced in Italy as an educational resources. It is open source, meaning that it can be manufactured and sold without paying royalties. It is available at Amazon.com for under $30. Just use a USB cable to power it up from any computer or USB power source and watch the LEDs light up. Touch a connected oscilloscope probe ground return lead to the USB housing and insert the needle point probe tip into the labeled terminals. You can turn the board over and probe circuit terminations. There are numerous sine wave variations, the DC power rails and lots of digital signals that can be displayed.

You will undoubtedly want to view utility power from residential or industrial outlets. Here great care should be taken, as previously emphasized, to avoid touching the ground return lead to a floating voltage. Also, it is essential to observe voltage and Cat location limitations. Moreover, if you make a breakout box for accessing power from a branch-circuit receptacle, a safe design should be employed so that the exposed terminals cannot short out. Figure 9.14 shows a device that should be used only under the supervision of an electrician or technician trained in avoiding electrical shock and arc hazards.

Fig. 9.14 Fixture for
accessing utility waveform.
(Judith Howcroft)

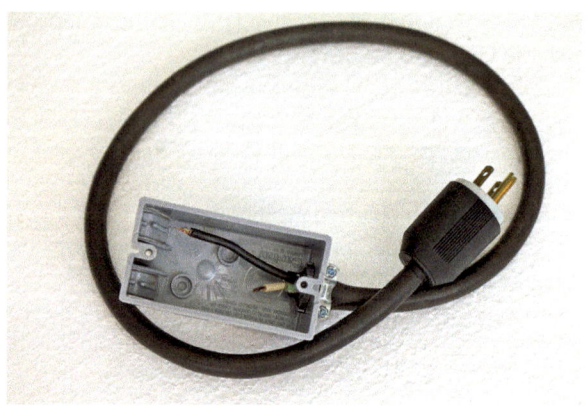

Chapter 10
What Lies Ahead?

Abstract Tektronix engineers speculate on future developments. Highest possible frequency as determined by Planck length, the smallest unit of spatial extension sets minimum wavelength. Waveform collapse as opposed to Many Worlds Interpretation of Hugh Everett.

Oscilloscope technology advanced rapidly from its early nineteenth-century roots through the triggered sweep and digital/flat-screen transformation in the 1980s. The trend has accelerated. Bandwidth and sampling rates have expanded exponentially. Additionally, new features and measuring and display capabilities have emerged. One wonders what lies ahead. The major oscilloscope manufacturers maintain robust research and development departments, with many theoreticians and engineers striving to introduce enhanced instruments.

To get a sense of where all this is going, we queried major oscilloscope manufacturers regarding innovations down the road. Among them was Tektronix. Gary Waldo, Product Planner, Core Real-Time Scopes, said: "Analysis will continue to expand beyond basic signal visualization. Most scopes in the midrange can acquire almost all signals of interest. But just seeing a signal isn't good enough anymore. Engineers want the scope to provide higher levels of information about the signal – whether its power analysis, decoding serial buses, signal integrity measurements, or much more advanced analysis like jitter".

"Scopes have become more upgradeable. In the past, bandwidths and record lengths had to be specified at the time of purchase. You got what you got. Now scopes are designed to be upgradeable, often in the field. This provides more flexibility to meet changing workloads and extends the life of the instrument. There will probably be further advances in the area of adding capacities of these instruments to meet whatever needs arise".

"Historically, scopes have been focused on the time domain. But some problems are easier to solve in the frequency domain, especially if you are able to tie your frequency domain events back to related time-domain phenomena. As such, we're seeing scopes move way beyond the traditional FFT by providing much more elegant and advanced frequency domain analysis with some scopes even providing an integrated spectrum analyzer. Tek has done some pioneering work on incorporating frequency domain analysis in scopes: for example, the mixed-domain oscilloscope".

D. Herres, *Oscilloscopes: A Manual for Students, Engineers, and Scientists*,
https://doi.org/10.1007/978-3-030-53885-9_10

235

"Usability is important. It may not sound like much, but if we can save the engineer a few seconds on each interaction with the scope by providing a simpler, more intuitive UI, that time really adds up. Many engineers use scopes in spurts, so it is important to minimize the "relearning curve." And beyond the time savings, the frustration level is kept down enabling them to focus on the problem at hand rather than focusing on how to drive the scope".

"We've recently introduced scopes that use a smart-phone-like interface. Scopes are precision instruments that have to measure thousandths of volts and billionths of seconds, so making them operate like consumer products is quite challenging. However, trends in consumer products will continue to find their way into oscilloscopes".

"Higher resolution and lower noise: Digital oscilloscopes have featured 8-bit vertical resolution for the last 15–20 years, as a lot of emphasis was put on speed. But signals have shrunk significantly from the days of TTL logic, to support much higher data rates. Power efficiency requirements and battery-powered devices are also leading to lower and lower-voltage signals. So recently scopes have progressed to 12-bit ADCs or even higher".

"This technology shift has allowed engineers to make more accurate measurements, especially on low-voltage signals. Of course, more vertical resolution only matters if you have a low-noise acquisition system that doesn't add noise to the measurement. Low noise enables you to bring the signal out of the noise and make a more accurate measurement."

John Marrinan, a UK Tektronix field application engineer, has this to say: What's the future for scopes? Your guess is as good as mine. I certainly think you will continue to see more integration of boxes on your lab bench into a single product, and I believe that this product will fundamentally be a scope. There will be more miniaturization and improved performance in modular-based scopes. And there will be a greater influence by the consumer on the test and measurements world. By this I mean smart-phone-type interfaces on scopes, instrument control through tablets, and more customization with widgets and apps.

"At the highest end of research things will stand still and there will be no change at all in my opinion…they'll just want higher bandwidths, faster sample rates and more accuracy like always."

While we're looking ahead, we should consider the future from a different perspective, not just that of instrumentation but looking at the electrical and electromagnetic waves themselves. Oscilloscope manufacturers have for years worked to realize greater bandwidth. This is an expensive undertaking, though quite doable. But have you wondered whether there is some maximum frequency that cannot in principle be exceeded? Frequency and wavelength are inversely related. It has been theorized that wavelength cannot be less than the Planck length, so this would mean an upper limit on frequency. Actual quantum graininess has been measured at orders of magnitude below the Planck length, which applies only to particles that have mass. This particular topic is speculative and controversial. Photons are said to have mass only if they are at rest. A moving photon, however, has energy, which is one aspect of mass.

A 2009 paper, cited in arXiv.org titled *Testing Einstein's Special Relativity with Fermi's Short Hard Gamma Ray Burst GRB090510*, stated that gamma-ray bursts are the most powerful explosions in the universe and probe physics under extreme conditions. Gamma-ray bursts are divided into two classes, of short and long duration, thought to originate from different types of progenitor systems. The physics of their gamma-ray emission is still poorly known, over 40 years after their discovery, but it may be probed by their high-energy photons. The photon sets limits on a possible linear energy dependence of the propagation speed of photons (Lorentz-invariance violation) requiring for the first time a quantum-gravity mass scale significantly above the Planck mass.

Considering the Planck length as the smallest unit of spatial extension, a wave having that length would exhibit a frequency of 6.2×10^{34} Hz. By way of comparison, a typical gamma-ray frequency could exceed a mere 10^{19} Hz, still many orders of magnitude slower and yet well beyond the bandwidth of our highest-priced oscilloscopes. Moreover, because energy and frequency of a photon are proportional, the highest frequency as set by the smallest wavelength would require vast amounts of energy, depending upon the actual amount of radiation involved. Just a single photon would have to measure 41 J of energy, which is 2.56×10^{20} eV.

Besides measuring and displaying extremely high-frequency waveforms, oscilloscopes of the future may play a role in the development and subsequent operation of artificial intelligence and of the universe itself. This need not be a hardware-based instrument. The software-based oscilloscope is currently in an early stage of development. Most models do not really compete with bench models. But in a post human-dominated world, artificially intelligent entities would undoubtedly have the ability to design and install software in their enormous computer networks that would be capable of realizing virtual oscilloscope instrumentation that could synthesize and modify what Hugh Everett termed the universal waveform in his many-worlds interpretation of quantum reality.

Hugh Everett was an eccentric individual and so were his ideas. But they were espoused by no less a respected and venerable theoretician as Stephen Hawking. Whether alternate realities exist and can be detected and even modified by our instrumentation is not known at present and may be in principle unknowable.

The many-worlds theory is an interpretation of quantum theory or quantum mechanics, as it has been known since shortly after it was conceived in 1925 by Werner Heisenberg. It states that the universal wave function is real. Upon observation and measurement, it collapses. This explains the dual nature of light and other particles in the double-slit experiment.

What Is Wave Function Collapse?

The quantum state of any classical body or system has an associated wave function. Since waveforms can be added, as by an oscilloscope in the Math mode, it follows that theoretically there is universal waveform of all things. In quantum theory, it was proposed that any wave function would collapse upon observation, reverting to its potential existence as an attribute of the body, which would remain unchanged. Collapse, then, is a reduction to zero. It is important to acknowledge that it is the wave function only, and not the body that collapses.

The many-worlds interpretation denies wave function collapse. The implication is that all possible outcomes of quantum measurements exist somewhere, which is to say in another world or universe. Measurements obviously can be just as well chosen and completed by machine or even by some natural process, and for this reason, there must be a vast number of alternate realities. Indeed, if an infinite number of possible realities does exist somewhere, it is not likely to be in our observable universe, or we would have noticed. If they are infinite, they would not fit.

This raises the possibility that these many vast systems must exist in numerous universes that may in principle be unknowable by us. An alternate hypothesis, at once equally outrageous and plausible, is that a future technology will permit us to detect these hidden worlds. A further implication is that sometime in the future (the way things are accelerating, it may be near rather than distant), we will develop instrumentation that will pierce these spatial and temporal barriers. If so, that most versatile of instruments, the oscilloscope, may be the key to it all.

More Information

Appendix A: Questions and Answers

Chapter 1

1. A waveform is

    (A) The actual electrical wave
    (B) A graph of an electrical or other wave
    (C) A physical entity traveling through space
    (D) A static picture of a wave in time
    Answer: B

2. The X- and Y-axes in an oscilloscope time-domain display are

    (A) Parallel lines representing time and amplitude
    (B) Parallel lines representing frequency and amplitude
    (C) Perpendicular lines representing frequency and amplitude
    (D) Perpendicular lines representing time and amplitude
    Answer: D

3. The X- and Y-axes in an oscilloscope frequency domain display are

    (A) Parallel lines representing time and amplitude
    (B) Parallel lines representing frequency and amplitude
    (C) Perpendicular lines representing frequency and amplitude
    (D) Perpendicular lines representing time and amplitude
    Answer: C

4. In the frequency domain in an oscilloscope display

    (A) Amplitude is shown on the X-axis as volts.
    (B) Amplitude is shown on the Y-axis as power.
    (C) Amplitude is shown on the X-axis as power.

© Springer Nature Switzerland AG 2020

D. Herres, *Oscilloscopes: A Manual for Students, Engineers, and Scientists*,
https://doi.org/10.1007/978-3-030-53885-9

(D) Amplitude is shown on the Y-axis as volts.
Answer: B

5. In the oscilloscope spectrogram mode

(A) Amplitude is shown as vertical motion.
(B) Amplitude is shown as various colors.
(C) Frequency is shown as vertical motion.
(D) Frequency is shown as various colors.
Answer: B

6. James Clerk Maxwell, in the nineteenth century, concluded that light was a form of electromagnetic radiation

(A) Because both traveled through a vacuum at the same speed
(B) Because both could be bent by a prism
(C) Because both were forms of energy
(D) Because both had waveforms that looked the same in an oscilloscope
Answer: A

7. The Michelson-Morley experiment

(A) Demonstrated the existence of the luminiferous aether
(B) Was inconclusive and baffled the researchers
(C) Remains a mystery to this day
(D) Was shown by Einstein to be based on faulty equipment
Answer: B

8. The LCD flat screen

(A) Is illuminated from behind by LED lighting
(B) Is considered obsolete
(C) Has a more powerful gun than the CRT
(D) Requires caution due to higher voltages than the CRT
Answer: A

9. Triggered sweep

(A) Is no longer used
(B) Was developed for use in the oscilloscope by Howard Vollum and colleagues at Tektronix
(C) Is lost whenever 10 percent noise is added to a signal
(D) Always begins at the same point on a positive-going slope
Answer: B

10. Modern oscilloscopes can toggle between time domain and frequency domain using

(A) Fast Fourier transform (FFT)
(B) Fourier analysis
(C) Fourier transform

(D) Energy from oscillating waves
Answer: A

Chapter 2

1. The simplest semiconductor is

   (A) A transistor
   (B) An FET
   (C) A MOSFET
   (D) A diode
   Answer: D

2. A diode is said to conduct when

   (A) It is not biased.
   (B) It is forward-biased.
   (C) It is reverse biased.
   (D) An external voltage is applied.
   Answer: B

3. Silicon has relatively few electrons because

   (A) It has a low valence number.
   (B) It has a high valence number.
   (C) The atoms are closely packed in a crystal lattice.
   (D) It is a perfect insulator.
   Answer: C

4. Bipolar junction transistors have

   (A) Two terminals and three layers
   (B) Three terminals and three layers
   (C) Three terminals and two layers
   (D) Two terminals and two layers
   Answer: B

5. In an FET

   (A) There is electrical current associated with the gate.
   (B) There is an electrical field associated with the gate.
   (C) The gate is analogous to the BJT emitter.
   (D) The gate is analogous to the BJT collector.
   Answer: B

6. A MOSFET schematic has

   (A) Two parallel lines with a space between them
   (B) A base, emitter, and collector

    (C) Two terminals

    (D) A source but no drain
       Answer: A

7. MOSFET's are

    (A) Immune to static charge

    (B) Highly vulnerable to static charge

    (C) Brittle and break easily

    (D) No longer used
       Answer: B

8. Lissajous patterns

    (A) Are useless curiosities

    (B) Require four oscilloscope inputs

    (C) May be displayed in an oscilloscope by pressing XY triggering

    (D) Are inherently unstable and must be constantly refreshed in an oscilloscope
       Answer: C

9. The BNC connector

    (A) Is easier to connect than a coax connector

    (B) Has poor frequency response compared to an oscilloscope probe

    (C) Is insulated from the oscilloscope chassis ground

    (D) Is used in all electronic equipment
       Answer: A

10. In an oscilloscope, AC coupling

    (A) Permits both AC and DC signal components to be displayed

    (B) Eliminates the DC component by inserting a capacitor in series with the signal

    (C) Eliminates the DC component by inserting a capacitor in parallel with the signal

    (D) Will not detect ripple in the output of a DC power supply
       Answer: B

Chapter 3

1. For a motor to work

    (A) There must be two magnetic fields, one rotating and one stationary or rotating at a different rate.

    (B) External power is not required.

    (C) Permanent magnets in the rotor won't work.

    (D) Permanent magnets in both rotor and stator will work.
       Answer: A

2. The purpose of the brush-commutator assembly in a DC motor is

   (A) Solely to introduce current into the rotor
   (B) Solely to reverse polarity periodically
   (C) To introduce current into the rotor and reverse polarity as required
   (D) To extract power for external use
   Answer: C

3. In a VFD, the equipment grounding conductor

   (A) Is not required
   (B) Must contact either L1, L2, or L3
   (C) Is continuous back to the electrical service
   (D) Must be copper
   Answer: C

4. In a VFD

   (A) There are two sections, rectifier and inverter.
   (B) There are two sections, rectifier and DC bus.
   (C) There are two sections, DC bus and inverter.
   (D) There are three sections, rectifier, DC bus, and inverter.
   Answer: D

5. In a VFD, the DC bus voltage

   (A) Is the same as the RMS AC line voltage
   (B) Is lower than the RMS AC line voltage
   (C) Is higher than the RMS AC line voltage
   (D) Should have over 10 percent AC ripple
   Answer: C

6. In a bench-type oscilloscope

   (A) Floating voltages can be measured by means of a differential probe.
   (B) The ground prong in the power plug should be removed.
   (C) Floating voltages are not hazardous.
   (D) Three-phase power can be displayed in all modes.
   Answer: A

7. In evaluating power quality

   (A) Harmonics are generally harmful.
   (B) Harmonics are most clearly displayed in the time domain.
   (C) Voltage among phases should not vary over 10 percent.
   (D) Current among phases should not vary over 3 percent.
   Answer: A

8. To service a CRT TV

   (A) Use a screwdriver to shunt out all dangerous voltages.
   (B) Use a power resistor to shunt out all dangerous voltages.

(C) An oscilloscope is not used.

(D) Soldering semiconductors does not require a heat sink as long as the iron does not contact the device.

Answer: B

9. Electrical noise

(A) Can be eliminated by use of acoustic padding

(B) Places a lower limit on any signal that can be displayed in an oscilloscope

(C) Cannot be generated within the oscilloscope

(D) Cannot be mitigated

Answer: B

10. Boltzmann's constant

(A) Relates the average kinetic energy of particles in a gas to the temperature of the gas

(B) Does not consider Nyquist noise

(C) Does not consider bandwidth

(D) Neglects current in amps

Answer: A

Chapter 4

1. To access the internal arbitrary function generator in an oscilloscope

(A) Power up the instrument, and it will appear by default.

(B) The easiest way is to insert a flash drive into the USB slot.

(C) Run a BNC cable from AFG Out to an analog channel input and press AFG.

(D) Turn all inputs off.

Answer: C

2. With AFG activated

(A) Use Multipurpose Knob a to select the desired waveform.

(B) Use Multipurpose Knob b to select the desired waveform.

(C) Press Math and toggle to select the desired waveform.

(D) After each waveform times out, the next one appears.

Answer: A

3. To display a waveform from another instrument

(A) A wireless connection is needed.

(B) The easiest way is to insert a flash drive into the USB slot.

(C) With the waveform displayed, run a BNC cable to an analog channel input and turn the channel on.

(D) With the waveform displayed, run a BNC cable to an analog channel input and turn the channel off.
Answer: C

4. In an oscilloscope AFG, arbitrary means

(A) They don't always work.
(B) Custom-made waveforms can be generated by the user.
(C) Waveforms are selected by a random-number generator.
(D) New waveforms are added over the Internet.
Answer: B

5. In an oscilloscope, to access Measure

(A) In Math, press Measure.
(B) In Acquire, press Measure.
(C) In Wave Inspector, press Measure.
(D) Turn Multipurpose Knob a until it comes up.
Answer: C

6. To access the DVM in an oscilloscope

(A) Press Measure in Wave Inspector and then DVM.
(B) Press Math and Acquire simultaneously.
(C) Press Default Setup and then turn Multipurpose Knob a.
(D) Press Default Setup and then turn Multipurpose Knob b.
Answer: A

7. Adding and subtracting frequency traces is done

(A) In the frequency domain
(B) In the time domain
(C) Using the DVM
(D) Using Wave Inspector
Answer: A

8. In the DVM, the user can access

(A) AC plus DC only
(B) DC only
(C) AC RMS only
(D) All of the above
Answer: D

9. Additional Measure submenus are

(A) Statistics only
(B) Gating only
(C) Reference Levels only
(D) All of the above
Answer: D

10. Automatic measurements are in

    (A) Time domain only
    (B) Frequency domain only
    (C) Time and frequency domain
    (D) Cursor mode only
        Answer: C

Chapter 5

1. Debugging begins by

    (A) Attempting to capture an infrequent event
    (B) Reducing the update rate
    (C) Compressing bandwidth
    (D) Adding more memory
        Answer A

2. Bandwidth is

    (A) Inversely proportional to frequency
    (B) Directly proportional to amplitude
    (C) A measure of the difference between highest and lowest frequencies within
        a continuous spectrum, generally in reference to a single unified signal
    (D) Not applicable when a signal modulates a carrier wave
        Answer: C

3. Attenuation at high frequencies is caused by

    (A) Matching characteristic impedance
    (B) Parallel capacitance
    (C) Series capacitance
    (D) Parallel inductance
        Answer: B

4. One way to reduce noise in an oscilloscope is to

    (A) Increase bandwidth.
    (B) Increase the number of channels.
    (C) Turn off waveform averaging.
    (D) Reduce bandwidth.
        Answer: D

5. High-resolution mode

    (A) Works only with real-time sampling
    (B) Works only with non-interpolated sampling
    (C) Works only with real-time, non-interpolated sampling

    (D) Displays a lower bandwidth waveform
        Answer: C

6. Average mode

    (A) Will not reduce noise in a signal
    (B) Uses sample mode for each individual acquisition
    (C) Takes an average for a random number of samples
    (D) Is less effective than bandwidth limiting
        Answer: B

7. Fast Acq

    (A) Can be toggled on and off.
    (B) When it is toggled on the waveform, palette submenu disappears.
    (C) Display is normal only.
    (D) Display is spectral only.
        Answer: A

8. Delay

    (A) Does not work with horizontal position
    (B) Is used in conjunction with horizontal position
    (C) Has no effect on the trigger point
    (D) Cannot acquire waveform detail
        Answer: B

9. LeCroy

    (A) Introduced triggered sweep.
    (B) Did away with the flat screen.
    (C) Made the oscilloscope obsolete.
    (D) Introduced the digital storage oscilloscope.
        Answer: D

10. Turning the horizontal scale knob clockwise

    (A) Increases the signal frequency
    (B) Reduces the signal frequency
    (C) Has no effect on the signal frequency
    (D) Increases the signal amplitude
        Answer: C

Chapter 6

1. The most common digital logic used currently is

    (A) Resistor-transistor logic
    (B) Diode-transistor logic

    (C) Transistor-transistor logic

    (D) CMOS-type MOSFET logic

       Answer: D

2. The semiconductor logic gate

    (A) Draws high current at its input

    (B) Draws low current at its input

    (C) Is a high impedance source at its output

    (D) Can tolerate only a single loading condition

       Answer: B

3. The oscilloscope is incapable of

    (A) Identifying infrequent and random glitches

    (B) Displaying digital and analog signals in a single screen

    (C) Detecting power supply anomalies

    (D) Repairing individual defective components

       Answer: D

4. Handheld multimeters

    (A) Are useless in testing complex circuits

    (B) Are capable of circuit measurements and diode checks in high-end instruments

    (C) Are rarely used these days

    (D) Cannot be used with floating voltages

       Answer: B

5. The logic probe

    (A) Is used only with an oscilloscope

    (B) Has LED'S indicating logic high, logic low, and pulse

    (C) Is very costly

    (D) Is no longer used

       Answer: B

6. The logic pulser

    (A) Injects CMOS or TTL logic high or logic low pulses into a circuit

    (B) Is used only with an oscilloscope

    (C) Is very costly

    (D) Is no longer used

       Answer: A

7. A logic analyzer

    (A) Is an inexpensive handheld instrument

    (B) Displays only a single channel

    (C) Is similar to a logic probe but with more extensive features

(D) Will not convert data into timing diagrams
Answer: C

8. To accurately convey data

(A) Sampling must be exactly equal to the Nyquist limit.
(B) Sampling must be less than the Nyquist limit.
(C) Sampling must be greater than the Nyquist limit.
(D) Sampling is independent of the Nyquist limit.
Answer: C

9. Serial data transmission

(A) Has largely replaced parallel data transmission
(B) Is applicable only to USB
(C) Is not used outside of PC's
(D) Is never bidirectional
Answer: A

10. CMOS circuits

(A) Use lots of power.
(B) Are susceptible to EMI.
(C) Produce EMI.
(D) Are used in limited numbers.
Answer: C

Chapter 7

1. When an IC fails

(A) It may be a short, which soon becomes an open circuit.
(B) It is usually because moisture enters the substrate.
(C) It is always visually apparent.
(D) The overall equipment is rarely affected.
Answer: A

2. In debugging a new design

(A) Test equipment is not used.
(B) The most versatile instrument is the oscilloscope.
(C) The multimeter is not used.
(D) Only the handheld oscilloscope is used.
Answer: B

3. Four-channel GHZ oscilloscopes

(A) Are needed for all troubleshooting
(B) Are found in most advanced shops and labs

    (C) Are not expensive

    (D) Have been replaced by handheld oscilloscopes

    Answer: B

4. The handheld, battery-powered oscilloscope

    (A) Will not read floating voltages

    (B) Has few advanced features

    (C) Is easily damaged in a factory setting

    (D) Is light and easy to move around

    Answer: D

5. The spectrum analyzer

    (A) Does measurements in the time domain

    (B) Does measurements in the frequency domain

    (C) Is less costly than an oscilloscope

    (D) Is not available as a PC-based instrument

    Answer: B

6. To measure amplitude and modulation of an RF signal

    (A) Use the network analyzer.

    (B) Use the signal analyzer.

    (C) Use the logic analyzer.

    (D) Use the protocol analyzer.

    Answer: B

7. The signal generator is

    (A) An AFG instrument only

    (B) An AWG instrument only

    (C) Either an AFG or an AWG

    (D) Neither an AFG nor an AWG

    Answer: C

8. Analog and digital circuitry

    (A) Are never used in a single piece of equipment.

    (B) Are combined in all electronic equipment.

    (C) Both exist in the digital storage oscilloscope.

    (D) Are not in the oscilloscope processor.

    Answer: C

9. In troubleshooting

    (A) A priority is to avoid introducing additional faults.

    (B) Speed is more important than accuracy.

    (C) Care must be taken when working with electrolytic capacitors because they are easily damaged.

(D) Audio equipment is not necessary to disconnect expensive speakers.
Answer: A

10. The signal generator is useful in radio receiver work because

(A) The tone is uniform, and frequency can be controlled.
(B) You can't blow the speaker.
(C) The instrument is inexpensive.
(D) The signal cannot be modulated.
Answer: A

Chapter 8

1. A quick and easy way to transfer a waveform from an oscilloscope to a computer is to

(A) Take a photograph of the screen.
(B) Connect to an existing telephone line.
(C) Insert a flash drive into the USB slot.
(D) Network via the Internet.
Answer: C

2. To connect an oscilloscope to a computer via LAN, begin by

(A) Connecting the oscilloscope to the modem via an Ethernet cable
(B) Removing all other Ethernet cables from the modem
(C) Turning off the modem
(D) Turning off the oscilloscope
Answer: A

3. Which of the following is true?

(A) Using a LAN connection, it is not possible to control the oscilloscope from the computer.
(B) Using a LAN connection, the successive screens are not dynamic.
(C) The LAN connection requires permission of the network administrator.
(D) The oscilloscope contains an internal server.
Answer: D

4. The LAN connection

(A) Requires that you get the oscilloscope IP address
(B) Has to be reconfigured if you turn off the oscilloscope
(C) Is inherently unstable
(D) Works anywhere in the world where there is Internet service
Answer: A

5. VISA

  (A) Is cross-platform.
  (B) Works with Windows-based PC's.
  (C) Drivers must be purchased.
  (D) Is wireless.
      Answer: B

6. LabVIEW

  (A) Is a product of National Instruments
  (B) Does not work on MAC computers
  (C) Does not work for data acquisition
  (D) Requires a USB cable
      Answer: A

7. MATLAB

  (A) Is a product of National Instruments
  (B) Is a product of MathWorks
  (C) Does not interact with Java
  (D) Does not include image recognition
      Answer: B

8. PING

  (A) Works only on Windows-based PC's.
  (B) Permits the user to perform diagnostic tests on a remote oscilloscope.
  (C) Does not require the oscilloscope's IP address.
  (D) Implementations in all utilities are the same.
      Answer: B

9. Ethernet

  (A) Is a frame-based technology.
  (B) Nodes all have the same MAC address.
  (C) Never used coax as a medium.
  (D) Has been replaced by token ring.
      Answer: A

10. An oscilloscope can display nonelectrical signals

  (A) When specialized drivers are installed
  (B) By means of appropriate transducers
  (C) With the exception of sound intensity
  (D) Including electromagnetic waves, by attaching directly to an antenna
      Answer: B

Chapter 9

1. PC-based oscilloscopes

   (A) Are simple instruments with none of the features found in bench-type models
   (B) Have ruggedized modules
   (C) Do not require software
   (D) Are available only in one-channel models
   Answer: B

2. PC-based oscilloscopes

   (A) Are more costly than handheld oscilloscopes.
   (B) Connect to a user-supplied computer
   (C) Do not require specialized software installed in the computer.
   (D) Consist of a module with exterior buttons and knobs.
   Answer: B

3. The computer

   (A) Is generally a PC, but Mac compatible modules are available.
   (B) Does not provide the display.
   (C) Does not provide the computing power.
   (D) Must be continually refreshed.
   Answer: A

4. After the waveform has been acquired by the computer

   (A) It can never be deleted.
   (B) It cannot be stored in the hard drive.
   (C) It can be copied into available applications.
   (D) It cannot be transferred to another computer.
   Answer: C

5. The PicoScope

   (A) Offers time and frequency domain modes
   (B) Ranges from $115 to $5000 depending upon bandwidth and number of channels
   (C) Has only one model
   (D) Is not available with AFG
   Answer: A

6. In the United States, the major oscilloscope manufacturers are

   (A) Tektronix, Teledyne Lecroy, and Keysight
   (B) PicoScope, Tektronix, and Agilent
   (C) PicoScope, Keysight, and LeCroy
   (D) Tektronix, PicoScope, and LeCroy
   Answer: A

7. Teledyne LeCroy

    (A) Is relatively inexpensive

    (B) Offers many advanced features

    (C) Is no longer manufactured

    (D) Does not offer cursors

       Answer: B

8. Tektronix

    (A) Is relatively inexpensive

    (B) Offers many advanced features

    (C) Is no longer manufactured

    (D) Does not offer cursors

       Answer: B

9. Keysight

    (A) Is relatively inexpensive

    (B) Offers many advanced features

    (C) Is no longer manufactured

    (D) Does not offer cursors

       Answer: B

10. Rigol

    (A) Manufactures high-quality, full-featured oscilloscopes with limited specifications at reduced cost

    (B) Does not manufacture multimeters

    (C) Does not manufacture spectrum generators

    (D) Does not manufacture arbitrary function generators

       Answer: A

Chapter 10

1. Oscilloscope inputs are

    (A) Oscilloscope probes only

    (B) BNC cables only

    (C) Neither probes nor cables

    (D) Both probes and cables

       Answer: D

2. For most work

    (A) The differential probe is preferred.

    (B) The 10:1 probe is suitable.

    (C) The 1:1 probe is suitable.

    (D) The current probe is needed.
      Answer: B

3. Active probes

    (A) Are less costly than passive probes.
    (B) Reduce reactive loading.
    (C) Are used at all times.
    (D) Do not contain semiconductors.
      Answer: B

4. Which of the following is true?

    (A) 1 GHz and lower oscilloscopes have Gaussian frequency response.
    (B) 1 GHz and higher oscilloscopes have Gaussian frequency response.
    (C) No oscilloscopes have Gaussian frequency response.
    (D) All oscilloscopes have Gaussian frequency response.
      Answer: A

5. The oscilloscope front end

    (A) Is usually disabled
    (B) Has a single signal path for all channels
    (C) Is defined as the paths from channel inputs to ADC
    (D) Scales the signals to the oscilloscope display
      Answer: C

6. The oscilloscope signal is digital

    (A) Beyond the ADC
    (B) And reverts to analog at the processor input
    (C) Prohibiting extensive processing
    (D) At the channel inputs
      Answer: A

7. In the oscilloscope digital section

    (A) There is one memory.
    (B) There are two memories.
    (C) There are three memories.
    (D) There are four memories.
      Answer: C

8. Techniques for increasing bandwidth include

    (A) Interleaving
    (B) DSP boosting
    (C) Raw hardware
    (D) All of these
      Answer: D

9. Interpolation consists of

   (A) Pulse
   (B) Line
   (C) Sin(x)/x
   (D) All of these
       Answer: D

10. The oscilloscope user

   (A) Cannot set record length
   (B) Can set record length
   (C) Cannot adjust memory depth
   (D) Cannot adjust the horizontal scale setting
       Answer: B

Appendix B: Electrical Laws and Equations

Ohm's law wheel, shown in Fig. AB.1, is a convenient circular tabulation that displays Ohm's law and the electrical power law. You can easily access solutions for volts, amps, ohms, and power unknowns, rather than performing multiple transpositions.

These electrical parameters are not empirical observations. They are fixed by definition. One volt, for example, is defined as the electrical potential that will cause 1 amp of current to flow through a 1-ohm load. These relations exist in the real world in accordance with the following definitions:

Fig. AB.1 Ohm's law wheel

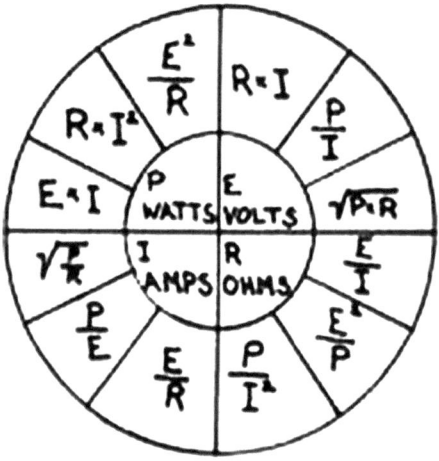

One amp is the amount of current that flows through a conductor when 6.24×10^{18} electrons pass a given point on the conductor each second. This is furthermore the definition of a charge of 1 coulomb.

Its clear that R refers to resistance in ohms, but what about E and I? E signifies electromagnetic force, the technical term for voltage. I signifies intensity, an old word for current. Voltage is applied to a load, and current flows through it. It makes no sense to say that 10 volts flows through a conductor or load or that 10 amps is applied to it.

The most basic equation, $E = I \times R$, is applicable when voltage is applied to and current flows through a purely resistive load. Not all loads are purely resistive. Some loads are capacitive, and some are inductive. Many (theoretically all) are partly resistive and partly capacitive or inductive. Capacitance and inductance are properties of a component, circuit, or transmission line, and they do not change unless some physical change takes place. In other words, a component will have a specific capacitance or inductance whether it is in packaging on a shelf or in an energized circuit.

Based on its capacitance or inductance, these components have capacitive and inductive reactance, measured in ohms. However, they do not have a fixed relationship to the amount of current as does resistance in a purely resistive load. Instead, capacitive and inductive are frequency dependent. At higher frequencies given the same current, capacitive reactance is less, and inductive reactance is greater. (To calculate the effect of these components, of course, you also have to look at whether they are connected in series or parallel to the load.)

The equation that governs capacitive reactance is:

$$X_C = 1 / 2\pi f C$$

where

X_C = capacitive reactance
f = frequency
C = capacitance

Since frequency is in the denominator, capacitive reactance varies with it. It goes down as frequency goes up.

The equation for inductive reactance is:

$$X_L = 2\pi f L$$

where

X_L = inductive reactance
f = frequency
L = inductance

Since f is no longer in the denominator, a rise in frequency causes the inductive reactance to go up.

Online Resources

Major oscilloscope manufacturers maintain extensive websites. There, along with extensive technical information and product specifications, you will find comprehensive information on available instrumentation, including oscilloscopes, spectrum analyzers, arbitrary function generators, and the like. Users' manuals are offered as free downloads, regardless of whether the instruments have been previously purchased. Students and technicians can acquire extensive electronic or print libraries of this material for reference as needed.

Index

© Springer Nature Switzerland AG 2020   259
D. Herres, *Oscilloscopes: A Manual for Students, Engineers, and Scientists*,
https://doi.org/10.1007/978-3-030-53885-9